CABI CONCISE

MEDICINAL AND AROMATIC PLANTS

Aromatic plants and spices are probably the most underestimated food worldwide. Although they have constantly accompanied humans in the course of evolution (co-evolution), they are only rarely found in official dietary recommendations. Yet they are not only aromatic, but extremely effective and can be used for everything from increasing well-being to treating various illnesses. They are, so to speak, the 'tastiest medicine in the world'. The essential oils contained in most spices, for example, not only give the spice its specific flavour, but also have various antibacterial, antiviral and anti-inflammatory properties.

There is a very long tradition of using herbs and spices for health. In traditional medicinal systems such as Ayurvedic, Chinese European Monastic, and Persian medicine, spices have been used to treat various diseases for more than 3000 years. Today, traditional medicinal systems are a topic of global importance. In many countries, a large proportion of the population relies heavily on traditional practitioners and medicinal plants to meet the need for basic medical care. This is because modern synthetic preparations are often only available to a limited extent and are usually very expensive. Medicinal and aromatic plants (MAPs), or herbal medicines, on the other hand, have often retained their popularity for historical and cultural reasons.

In recent decades, increasing attention has been paid to the phytochemical and pharmacological study of traditional medicinal plants in order to discover and develop new, safe and affordable medicines. These medicinal plants are not only derived from the well-known traditional medical systems, but also refer to the experiential knowledge of indigenous people and tribes in South America and Africa. For example, the root of devil's claw (*Harpagophytum procumbens*) was used extensively for healing purposes by the first inhabitants of South Africa.

Since human olfactory cells have direct access to the oldest regions of our brain, such as the limbic system and the hippocampus, where emotions, feelings and, above all, our memories are stored, scents have a direct effect on our well-being. It therefore seems worthwhile to consider how traditional knowledge about the health benefits of various medicinal and aroma plants can be re-evaluated using the techniques available today, and how the results can be utilized for the pharmaceutical and food sectors.

This series focuses on traditional medicine systems and covers different medicinal plant species or products. Each book will largely follow the same content structure, and will cover the history, efficacy, active ingredients, occurrence, genetics/breeding, cultivation and economic importance.

As the market demand for natural ingredients is increasing, this series aims to provide succinct and readable overviews of key medicinal and aromatic plant species used around the world. We welcome new proposals and suggestions.

Hartwig Schulz

Titles Published

Essential Oils Unveiled: Complex Compositions for Food, Cosmetics, and Medicine
Marek Bunse, Hartwig Schulz et al.

Saffron: Bridging Traditional and Modern Applications in Nutrition and Medicine
Hassan Rezadoost, Hassan Esmaeili, Edris Mahdavi-Fikjvar and Hartwig Schulz

Cardamom: Advances in Breeding, Production and Quality
R. Sivaranjani and H.J. Akshitha

Titles in Preparation

Frankincense: Applications in Traditional and Modern Medicine
Hidayat Hussain

Cardamom

Advances in Breeding, Production and Quality

Edited by

R. Sivaranjani

ICAR-Indian Institute of Spices Research, Kozhikode, Kerala

H.J. Akshitha

ICAR- Indian Institute of Spices Research, Regional Station, Appangala, Coorg, Karnataka

CABI

CABI is a trading name of CAB International

CABI
Nosworthy Way
Wallingford
Oxfordshire OX10 8DE
UK

Tel: +44 (0)1491 832111
E-mail: info@cabi.org
Website: www.cabi.org

CABI
200 Portland Street
Boston
MA 02114
USA

T: +1 (617)682-9015
E-mail: cabi-nao@cabi.org

A catalogue record for this book is available from the British Library, London, UK.

ISBN-13: 9781836995814 (hardback)
9781836995821 (paperback)
9781836995838 (ePDF)
9781836995845 (ePub)

DOI: 10.1079/9781836995845.0000

Commissioning Editor: Rebecca Stubbs
Editorial Assistant: Theresa Regueira
Production Editor: Theresa Regueira

Typeset by Exeter Premedia Services Pvt Ltd, Chennai, India
Printed and bound in the UK by CPI Group (UK) Ltd, Croydon, CR0 4YY

Contents

Preface

Cardamom is among the most valuable and distinctive spice crops in the world, prized for its complex aroma, culinary versatility, and cultural significance. For centuries, it has played an important role in traditional medicine, gastronomy, and international trade, linking regions of origin with global markets. Today, cardamom cultivation supports the livelihoods of millions of smallholder farmers especially in India while serving a steadily growing demand from food, beverage, pharmaceutical, and cosmetic industries.

Despite its high value, cardamom production faces a range of interconnected challenges. Climate change, increasing pest and disease pressure, limited availability of improved planting material, and rising production costs all threaten productivity and sustainability. At the same time, quality requirements have become more stringent, driven by food safety regulations, traceability demands, and evolving individual consumer preferences. Market structures are also changing, influenced by price volatility, shifting trade flows, and the need for greater transparency and value addition along the whole supply chain.

This book provides a comprehensive overview of the topics mentioned by bringing together knowledge from various disciplines. The chapters deal with genetic resources and breeding strategies, agronomic practices and production systems, post-harvest processes and quality assessment, as well as market dynamics and economic considerations.

The book is aimed at scientists, consultants, students, policy makers and stakeholders in the spice industry who are seeking an in-depth understanding of cardamom and its future prospects.

Hartwig Schulz
(hs.consulting.map@t-online.de)

Contributors

H.J. Akshitha, ICAR-Indian Institute of Spices Research, Regional Station, Appangala, Coorg., Karnataka, India

M. Alagupalamuthirsolai, ICAR-Sugarcane Breeding Institute, Coimbatore, Tamil Nadu, India

S.J. Ankegowda, ICAR-Indian Institute of Spices Research, Regional Station, Appangala, Madikeri 571201, India

Honnappa Asangi, ICAR-Indian Institute of Spices Research, Regional Station, Appangala, Madikeri, Coorg (D) – 571201, Karnataka, India

Kaliyaperumal Ashokkumar, School of Agriculture and Animal Sciences, Gandhigram Rural Institute-Deemed to be University, Gandhigram, Dindigul, Tamil Nadu, India

K. Devikrishnan, Centre for Textual Studies and Publications, Arya Vaidya Sala, Kottakkal, Kerala, India

C.M. Harinarayanan, Centre for Medicinal Plants Research, Arya Vaidya Sala, Kottakkal, Kerala, India

A. Indhushree, Department of Agricultural Economics, SRM College of Agricultural Sciences, SRM Institute of Science and Technology, Chengalpattu – 603201, India

Nimisha Mathews, Cardamom Research Station, Pampadumpara, Kerala Agricultural University, Idukki, India

M. Murugan, Cardamom Research Station, Kerala Agricultural University, Pampadumpara, Kerala, India

Saradhadevi Muthukrishnan, Department of Biochemistry, Bharathiar University, Coimbatore, Tamil Nadu, India

G.A. Rajanna, ICAR-Indian Institute of Spices Research, Regional Station, Appangala, Madikeri 571201, India

M.S. Shivakumar, ICAR-Indian Institute of Spices Research, Regional Station, Appangala, Madikeri, Coorg (D) – 571201, Karnataka, India

R. Sivaranjani, ICAR-Indian Institute of Spices Research, Kozhikode, Kerala, India

H.C. Vikram, Pepper Research Station, Kerala Agricultural University, Panniyur, Kannur 670142, India

Introduction

1

R. Sivaranjani[1]* and H.J. Akshitha[2]

[1]*ICAR-Indian Institute of Spices Research, Kozhikode, Kerala, India*
[2]*ICAR-Indian Institute of Spices Research, Regional Station, Appangala, Coorg., Karnataka, India*

Cardamom, regarded as the 'Queen of Spices' contains the flavour that overpowers even the strongest perfume. Owing to its unique flavour profile, it is irreplaceable by any other aromatic crops. Cardamom is native to the Indian subcontinent and originated in the Western Ghats regions of India comprising present-day Kerala, Karnataka and Tamil Nadu. It is scientifically called *Elettaria cardamomum* (L.) Maton, the name is an amalgamation of its local name 'Elam', 'Elakki' and 'Elakkai' in the Dravidian languages and its Latin word 'cardamōmum'. For centuries, its green pods filled with aromatic seeds were used for flavouring, healing and in ancient rituals across the world.

If we look back into the history, worlds were discovered, wars were fought and countries were colonized for trading Indian spices by European traders – turned conquerors. The multiple spectra of history, geography, politics and economics converge in the lives of labourers and owners of cardamom plantations in famous cardamom hills of Kerala where huge plantations were established during the colonial period. The Western Ghats area where cardamom had evolved naturally was inhabited entirely by tribal population in the 1800s and cardamom was collected from the wild and considered as minor forest produce. Slowly, small-scale gardens of cardamom were established in the hills by local people from adjacent areas of Bodinaikannur and Cumbum near the cardamom hills and these areas formed the primary markets for cardamom trade.

Between 1800 and 1880, due to the depletion of forest resources caused by high demand, the plantations slowly started emerging, encouraged and supported by local rulers, although the ownership of the land was always under contention between Travancore government and local ethnic Tamils.

*Corresponding author: ranjanigop@gmail.com

The period spanning 1880–1940 was characterized by the establishment of large-scale plantations on the line of tea and coffee plantations established by European settlers. The local rulers followed the European pattern of establishing cardamom plantations on the land controlled by them. The European traders started their negotiations with the local rulers from both Tamil Pandya and Malayali kingdoms for the access and trade of cardamom grown in the hills through several deeds signed between 1861 and 1890. Lured by the prospects of cardamom trade and generation of revenue, the Travancore government slowly and cautiously started granting rights to develop cardamom plantations in the hills to European planters. Slowly, collectors of cardamom were turned into cultivators and granting of occupancy rights by the government was emerged to prevent the illegal trade on the produce (Moench, 1991).

The study '*The cardamom conundrum resolved*' (Poulsen *et al.*, 2024) redefined the genus *Elettaria* within the ginger family (Zingiberaceae) using morphological and molecular evidence. It clarified the genus's placement in the only pantropically distributed ginger lineage and recognized seven species, including two new ones from the Western Ghats of India (*E. facifera* and *E. tulipifera*). Several Southeast Asian species were excluded from *Elettaria*. This recircumscription resolves long-standing taxonomic confusion and enhances understanding of cardamom evolution and biogeography.

From its centre of origin, the crop was introduced to Southeast Asian nations such as Malaysia, Indonesia, Thailand, Cambodia and Vietnam due to the favourable environmental conditions for its growth and reproduction. In 1920, it was introduced in Guatemala and the country presently surpassed India as the top cardamom producer and exporter in the world. Now, only a handful of countries produce cardamom for global demand.

Cardamom has a rich history of culinary usage among global cuisines. Evidence from ayurvedic texts such as Charaka Samhita and Sushruta Samhita suggests the use of cardamom as far back as 1500–1000 BC (Mahindru, 1982). In ancient India, cardamom was considered a luxury spice, used in religious rituals, perfumes and as a status symbol in culinary preparations. It is the cornerstone of Indian cooking, used profusely in both sweet and savoury dishes such as kheer, gulab jamun, curries, biryani and pulao. It also plays a pivotal role in Middle Eastern, Persian and Afghan cuisines, used exclusively for its flavour and fragrance. It was also utilized for its medicinal properties to alleviate the symptoms of digestive issues and respiratory ailments.

Cardamom requires a shaded, humid environment with well-drained, fertile soil (pH 6.0–6.8) and temperature of 15–35°C. Plants are propagated via seeds or suckers. Seedlings are raised in nurseries for almost a year before transplanting. The productive lifespan of cardamom is for 10–15 years, with economic yield starting after 3 years. It is a perennial herbaceous rhizomatous crop with the average height of 1.5–3 m and canopy diameter of 1.2–1.5 m. Cardamom is a shade-loving crop, requiring 40–60% shade level for maximum growth and yield. Sometimes a reduction of up to 60% yield is noticed in plants grown in unshaded areas. In India, sustainable farming practices promote intercropping cardamom with coffee and native trees for optimum shade requirement. Cardamom is a nutrient-intensive crop that requires both organic and inorganic fertilizers for tillering, root development, optimum growth, yield and quality (Prabhakaran Nair, 2006). The soil-based nutrient application for targeted yield introduced by ICAR - Indian Council of Agricultural Research, Kozhikode, Kerala, is becoming increasingly important for cardamom growers along with good agricultural practices for sustainable yields (Ankegowda *et al.*, 2022b).

The inflorescence emerges from the base of the tiller (Fig. 1.1a, b), and the period from the initial swelling of the reproductive bud to full panicle development lasts about 6–7 months. Inflorescence emergence typically begins in February–March following the pre-monsoon showers, with peak panicle initiation occurring during the monsoon months (June–August). Flowering in cardamom is non-synchronous and occurs concurrently with inflorescence development. It takes place during the panicle elongation stage and continues for about 4–5 months. The first flower in a panicle opens approximately 60–80 days after panicle initiation, with peak flowering observed during June–August (Akshitha *et al.*, 2024).

The fruit of cardamom is called capsule (Fig. 1.1c, d). Botanically, it is a trilocular, dry, dehiscent fruit formed from a superior ovary with three carpels each forming a single locule, splits along septa at maturity. The outer

Fig. 1.1. (a) Inflorescence emergence from the base of the tiller; (b) Cardamom flower; (c) Fresh cardamom capsule; (d) Dry cardamom capsule; (e) Cardamom fresh seeds with mucilage.

layer covering the capsule is pericarp, which encloses the locule containing 6–18 seeds. The seeds are kidney shaped, brown–black coloured and attached via aril to the capsule (Fig. 1.1e). The varieties are categorized into Malabar, Mysore and *Vazhukka* types based on panicle type, minor variations in shape and size of capsules, but the core structure remains the same. The Malabar type is spread across different parts of Karnataka whereas the Mysore type is mainly cultivated in Kerala and Tamil Nadu. The growth of the panicle in the Vazhukka type exhibits intermediate characters of the Malabar and Mysore types and is considered to be a natural hybrid between these two types.

Over the years, different genotypes (varieties) obtained through selection and hybridization were released from universities and research organizations across the country. Organic amendments like compost or manure are preferred in sustainable systems. Integrated nutrient management uses cover crops (e.g. legumes) to fix nitrogen. Soil erosion control involves terracing on slopes. The rainfed cultivation of cardamom requires well-distributed rainfall of 1500–2500 mm/year, but microirrigations such as overhead sprinkler, mini sprinklers and mist irrigation can be adopted to mitigate the drought period associated during summer months.

The cardamom capsules ripen at 120–135 days after pollination. In India, the harvesting begins in the month of July–August and lasts until January–February with an interval of 15–20 days between harvests. Owing to its staggered maturity, harvesting is done manually with workers picking the ripe capsules and leaving the unripe capsules behind, which requires skill and training. The harvested capsules are washed with water to remove dirt and other adhering materials and dried in traditional firewood oven or mechanical convection oven for 18–24 h to obtain moisture levels of 12–15%. Natural sun drying is generally avoided as it leads to loss in colour and marketability of the produce.

In India, among different states, Kerala contributes to around 70% of national production followed by Tamil Nadu. During 2023–2024, in India, cardamom was cultivated in an area of 70,410 ha with a production of 25,230 t (Spices Board of India). In 2024–2025, India exported 6728 t of small cardamom valued at Rs. 1566.82 crores. Key importers of Indian cardamom are Saudi Arabia, UAE, Kuwait, Qatar, USA, Europe, Bangladesh, Pakistan, Japan and North Africa. Guatemala's cardamom production recorded a 127.43% increase over the past 5 years, rising from 41,864.74 t in 2019 to 95,210.95 t in 2023. As of 2023, Guatemala ranked first globally in cardamom exports, accounting for approximately 53.28% of the total global export value, with exports valued at US$365,005,733. The major importers of Guatemalan cardamom include Saudi Arabia, the United Arab Emirates, Bangladesh and Pakistan.

The cardamom capsules are graded according to key parameters such as capsule size, colour, texture, moisture content, essential oil content, seed count, cleanliness and defects. In India, the grading system and auctioning is governed by the Spices Board ensuring the quality of the produce for domestic and export markets. Two main commercial grades in India are Alleppey Green and Coorg Green, which is further categorized into Bold, Super Bold and Extra Bold based on the size of the cardamom capsules. The Guatemalan cardamom is graded as Jumbo Green, Imperial Green, Fancy Green and Mixed Green which are further classified into different names based on the size and colour of the capsules.

The unique flavour of the cardamom is due to the presence of essential oil constituents such as 1,8 cineole and alpha-terpinyl acetate present in the seeds. Apart from these two compounds, other major volatile constituents of the cardamom essential oil include alpha-pinene, myrcene, linalool, limonene, terpinene, geraniol, etc. Generally, there is no significant difference in the volatile profile of Indian cardamom and Guatemalan cardamom (Ashokkumar *et al.*, 2021; Kiralan *et al.*, 2023).

As compared to Guatemala, cardamom is a highly domestic market oriented in India. In 2023, India produced around 18,310 t of cardamom, out of which 6168 t (34.02%) worth US$120.74 million were exported. (https://www.indianspices.com/sites/default/files/major%20itemwise%20export%202025%20web.pdf). United Arab Emirates (UAE), Saudi Arabia and USA are major importers of Indian cardamom. In the same year, Guatemala produced 95,210 t of cardamom, out of which 52,691 t (55.34%) were exported. Around 61% of cardamom imported by Europe is sourced from Guatemala. (https://wits.worldbank.org/trade/comtrade/en/country/IND/year/2023/tradeflow/Exports/partner/ALL/product/090830)

Cardamom is a labour- and input-intensive crop and due to this, the plantation faces several challenges such as increasing production cost due to labour shortages and pesticide residues due to increased application to protect the crop from pests and diseases. Furthermore, increasing price fluctuations on the national and international markets are leading to cardamom plantations gradually being replaced by other crops such as coffee or fruits. The government of India through its agencies such as Spices Board and ICAR-Indian Institute of Spices Research undertaking several initiatives to supply quality planting material, advanced technologies for increasing production and protect the plants from pests and diseases, which may ultimately lead to increase in national production and to increase the share of Indian cardamom in the world market.

Cardamom (*Ela*) in Ayurveda

2

K. Devikrishnan[1] and C.M. Harinarayanan[2]*

[1]*Centre for Textual Studies and Publications, Arya Vaidya Sala, Kottakkal, Kerala, India*
[2]*Centre for Medicinal Plants Research, Arya Vaidya Sala, Kottakkal, Kerala, India*

2.1 Cardamom in Ayurveda

Cardamom, known as the 'queen of spices', belongs to the family Zingiberaceae. It's a delightful spice derived from the seeds of the perennial plant, *Elettaria cardamomum* Maton. In Ayurveda, it is termed as *Ela*. It's been celebrated for its medicinal properties for ages and has even found mentions in Greek, Egyptian and Christian writings. In India, the best cardamom comes from the lush high ranges of Idukki in Kerala, as well as some areas in Karnataka and Tamil Nadu. It's mostly known for its pungent taste, with a sweet note that follows, making it a wonderful addition to any dish. What's really great is that *Ela* helps keep the body's three doshas – *vata*, *pitta* and *kapha* – balanced, which is why it's cherished not just for flavour but for well-being too. Cardamom's distinctive aroma is attributed to the volatile oils found within its seeds and capsules. Historically, cardamom has been incorporated with betel leaves for chewing, valued for its ability to purify the mouth. The seeds of cardamom are characterized by their aromatic and cooling properties, and are known for being stimulant, carminative, digestive, stomachic, diuretic, cardiotonic, alexiteric and expectorant. These properties make cardamom effective in addressing a variety of conditions, including anorexia, dyspepsia, asthma, bronchitis, dysuria, renal and vesicle calculi, halitosis, cardiac disorders, hyperdipsia, burning sensations, debility and vision defects. In the Ayurvedic system of medicine, cardamom is used in the preparation of medicines for applications including treating infections of the teeth and gums, addressing throat ailments, alleviating lung congestion, managing pulmonary tuberculosis and reducing inflammation of the eyelids. Additionally, cardamom offers solutions for gastrointestinal disorders, promoting kidney health and addressing issues related to impotency,

*Corresponding author: harinarayanan@aryavaidyasala.com

erectile dysfunction and premature ejaculation in Ayurveda. It is also used in the management of gall bladder stones among other conditions (Dattatray *et al.*, 2016; Singletary, 2022).

2.2 Historical Background

The mention about the usage of cardamom can be seen in the texts of the Vedic period. In the *Atharva Parishishta*, cardamom is mentioned as being used in the practice of *Nakshatra Snana*, which involves bathing at a specific time aligned with the lunar mansion. In the *Charaka Samhita*, cardamom is identified as a *Shirovirechaka Dravya* (a substance employed in purification therapy administered through the nasal passage). It is also classified as a *Vamanakaraka Dravya* (an emetic drug to induce vomiting for therapeutic purposes). Cardamom is referenced as a remedy for massaging at the *Raktamokshana* (a therapeutic procedure in Ayurveda, particularly when blood letting is improperly executed) site. *Vagbhata* also recognized cardamom as *Vamanakaraka Dravya*. This means cardamom is traditionally utilized to induce vomiting as part of therapeutic practices, aiding in the cleansing of the stomach by expelling unwanted substances (Tripathi, 2009).

2.3 Cardamom in Ayurveda Classics and Lexicons

The usage of cardamom (*ela*) has its reference in various ayurvedic *nighantus* (texts) where drugs are classified in different *Vargas* (groups) as per their properties. In *Dhanvantari Nighantu* and *Sodhala Nighantu*, cardamom has been mentioned in *Shatapushpadi Varga* (Sharma *et al.*, 1978; Sharma *et al.*, 2005). In *Shadarasa Nighantu*, cardamom has been mentioned in *Katudravyaskandha*. In *Madanapala Nighantu*, cardamom has been mentioned in *Karpuradi Varga*. In *Kaiyadev Nighantu*, cardamom has been mentioned in *Oshadhi Varga*. In *Bhavaprakasha Nighantu* also cardamom is in *Karpuradi Varga*. In *Rajanighantu*, cardamom has been included in *Pippalyadi Varga*. In *Mahaushadhi Nighantu*, cardamom has been mentioned in *Trijaat* and *Chaturjaat*. In *Priya Nighantu*, cardamom has been included in *Haritakyadi Varga*. In *Adarsh Nighantu*, cardamom has been included in *Aardrakadi Varga*. In *Gunaratnamala*, cardamom has been mentioned in *Karpuradi Sugandha Varga* (Tripathi, 2006).

The synonyms of this drug are *truti* (small seed), *chandra* (coolant), *chandrabala* (similar to moon), *tutta* (alleviate pricking pain), *triputa* (fruit triangular shaped), *korangi* (cardamom grows in South India), *upakunjika* (contracted fruits) and *kapotavarna* (having the colour of pigeon). Synonyms of cardamom from various available *nighaṇṭus* are furnished in Table 2.1.

Table 2.1. Synonyms of cardamom in different Ayurveda lexicons.

Names	DN	SN	SN	MN	KN	BPN	RN	PN	GR
Baalaa					*		*		
Baalaila					*				
Bahula	*	*		*					
Bahulagandhaa							*		
Balavatee							*		
Chandra					*				
Chandrabaala	*			*	*				
ChandravEla		*							
Chandrika							*		
Dravidee	*	*		*	*	*	*		*
Dwipa					*				
Eindree		*					*		
Elaa	*	*	*	*	*	*	*		
Gandhaphalikaa							*		
Garbhari							*		
Gaurangee							*		
Ghritaa		*							
Himaa							*		
Kapotavarna	*	*		*	*				
Kapotvarnee							*		
Kayastha							*		
Korangee	*	*			*	*			*
Kshudraila					*				
Kunatee				*					
Mashee					*				
Nishkutee	*	*		*	*		*		
Nistrutee					*				
Sagargaaminee							*		
Sugandha		*							
Sukshamaila		*		*	*	*		*	*
Sukshma		*	*		*	*	*		*
Sukshmadya			*						
Tridiva					*				
Trushti				*					
Trutee		*			*	*	*		*
Tuttha	*	*			*	*			*
Twisha				*					
Upakunchee							*		
Upakunchika		*			*	*			*
Vayastha		*							

BPN: Bhāvaprakāśa Nighaṇṭu, DN: Dhanvantari Nighaṇṭu, RN: Rājanighaṇṭu,
KN: Kaiyādeva Nighaṇṭu, MN: MadanapālaNighaṇṭu, PN: Priyanighaṇṭu, SN: Sodhāla Nighaṇṭu

2.4 Varieties and Substitutes

There are two varieties of this drug known as *sukshmaila* (*E. cardamomum*) and *Brihadaila* (*Ammomum subulatum*) in Ayurvedic texts. According to *Ashtangahridaya* it is included in *Eladigana*, which is a group of drugs used to improve the skin colour, in urticarial, itching and swelling. *Ela* (cardamom), *Tvak* (cinnamon) and *Patra* (bay leaves) are together known as *Trijatha* and *Trijatha* with *Nagakesara* (mesua) is called *Chaturjatha*. These combinations are ingredients in several Ayurvedic preparations such as *Lehas* (paste), *Arishtas* (fermented liquids) and *Churnas* (powders) (Grover, 2021).

2.5 Pharmacological Actions of Cardamom Mentioned in Ayurveda

The fragrant seeds are abundant in numerous essential vitamins, such as riboflavin, niacin and ascorbic acid, which are crucial for maintaining optimal health conditions. In the Ayurvedic system of medicine, these seeds are effectively utilized for various health issues, with significant applications including infections of the teeth and gums, throat ailments, lung congestion, pulmonary tuberculosis, eyelid inflammation, gastrointestinal disorders, kidney disintegration, impotence, erectile dysfunction, premature ejaculation and gall bladder stones (Korikanthimath *et al.*, 2000a).

The major therapeutic claims as per classical texts are, *Mutrakrichrahara* – diuretic, assisting individuals with dysuria and urinary retention; *Arshahara* – for the treatment of piles and haemorrhoids; *Shwasahara* – against asthma and chronic respiratory ailments; *Kasahara* – for alleviating coughs and colds; *Hridya* – as a cardiac tonic, promoting heart health; *Rochna* – enhancing taste and addressing anorexia; *Deepana* – boosts digestive processes; *Vatahara* – useful in countering neuralgia, paralysis, constipation and bloating; *Pittartihara* – provides relief from gastrointestinal pain and the sensation of burning (Grover, 2021). *Rasapanchaka* (five basic characteristics and *Karma* (therapeutic actions) of *Ela* from various *Nighaṇṭus* are tabulated in Table 2.2.

2.6 Usage in Ayurvedic Medicines

Cardamom is a versatile ingredient found in various traditional Ayurvedic formulations, each serving a specific therapeutic purpose (Table 2.3). It is incorporated into various ointments for treating skin diseases like *Kushtha*, as well as headaches and dental caries. In *Charaka Samhita*, it is classified within the *Shwasahara* and *Angamardaprashamana* groups and is also included in the *Katukaskandha* category in *Sushruta Samhita*. Cardamom is an ingredient in the preparation of *Chavanapraasa* and *Brihani Gutika*, which is renowned for its aphrodisiac properties. It is also part of *Agurvadya Taila*, used to treat cold fevers. Alongside other drugs, cardamom is mixed with soil to address

Table 2.2. Properties and actions of *Ela*.

Text	Rasa	Guṇa	Vīrya	Vipāka	Karma
DN	Madhura	Laghu	Sheeta		*Hridya* (cardiotonic), *Rochana* (decreases bitterness in mouth), *Deepana* (increases digestive fire), *Mukhashodhana* (mouth cleanser)
KN	Katu	Laghu	Sheeta	-	-
BPN	Katu	Laghu	Sheeta	-	-
RN	Tikta	-	Sheeta	-	-
PN	Katu	-	-	-	*Mukhashodhana* (mouth cleanser)
SN	Katu	Laghu	Sheeta	-	-
MN	Katu	Laghu	Ushna	Katu	-
GR	Katu	Laghu	Sheeta	-	-

Mritabhakshana, a behaviour involving the consumption of non-nutritive substances. In the treatment of bronchial asthma, cardamom is a component of *Prapaundarikadi Dhumapana*. It is also part of *Mritsanjivana* and *Mahagandhahasti Agada*, both used for treating various types of poisonings. Furthermore, cardamom is included in the dietary preparation *Mridvika Rasa* to enhance taste and stimulate digestion (Anonymous, 2011). As an ingredient in *Kampillakadi Taila*, it aids in wound healing. *Tejovhadi Churna*, which contains cardamom, is applied to gums to alleviate bleeding, pain and irritation. Cardamom is part of *Vyoshadi Modaka Yoga*, a formulation used in purgation therapy for conditions like dysuria and different forms of poisoning. It forms a crucial component of *Koshatakyadi Niruha Basti*, specifically administered to individuals suffering from ailments attributed to *Kapha Dosha* imbalances. In Ayurvedic treatment, cardamom is recognized as a *Prakshepa Dravya*, meaning it is added as an auxiliary substance, enhancing the effectiveness of various types of *Vasti* (enema) therapies (Anonymous, 2003).

In the context of *Raktamokshana*, a bloodletting procedure prominently featured by *Acharya Sushruta*, where cardamom is used as a supportive dietary component. It is also included in *Neelyadi Rechaka Churna*, a preparation used for treating conditions known as *Sannipatika Vyadhi*, where multiple *Doshas* are involved. Cardamom is also a component of *Gandha Taila*, employed in the management of fractures (*Bhagna*) and *Bhagandaranashana Taila*, used for addressing fistulas (*Bhagandara*). After childbirth, cardamom in combination with other medicinal substances is utilized for *Doshasyandana*, aiding the movement of *Doshas*, and *Vedanashamana*, providing pain relief. In *Bala Taila*, cardamom is used to treat a range of conditions associated with *Vata* disorders. Similarly, in *Vachadi Taila*, it is utilized for *Anuvasana Basti*, an oil enema procedure specifically targeting *Vata*-related ailments. Cardamom also

Table 2.3. Ayurvedic medicines containing cardamom.

SN	Formulation	Used in	Reference
1.	*Ajaswagandhadi Leham*	General and sexual debility, emaciation, deficient immune conditions, loss of libido and post-natal care	Proprietary medicine
2.	*Amritaprasaghritam*	Oligospermia, general and sexual debility, leucorrhoea and weight loss associated with chronic degenerative disorders	AH, Adhyayam 3
3.	*Aragwadharishtam*	Mild laxative and blood purifier. Effective in chronic ulcers, leucoderma and worm infestations	Proprietary medicine
4.	*Aranyatulasyadi Keratailam*	Used in atopic dermatitis	Proprietary medicine
5.	*Aravindasavam*	Improves immunity and thereby helps to control infectious diseases	BR, Adhyayam 17
6.	*Arimedadi Tailam*	Toothache, painful infection of teeth and gums can be used for Gandusham and Nasyam	SY
7.	*Asanamanjishthadi Tailam*	Prevents catarrh, head and neck diseases	Proprietary medicine
8.	*Asaneladi Tailam / Keratailam*	Prevents catarrh, head and neck diseases	Proprietary medicine
9.	*Aswagandhadi Ghritam*	In all cases of recurrent lung infection, emaciation, easy fatigability, sexual debility and prevention of premature aging	BR, Adhyayam 74,
10.	*Avipattichurnam*	A safe purgative in bilious diseases and habitual constipation	AH, Adhyayam 3
11.	*Ayaskriti*	Diabetes mellitus, obesity and anaemia	AH, Adhyayam 12
12.	*Balarishtam*	Vatavyadhis of varied aetiology	BR, Adhyayam 26
13.	*Balatailam*	Rheumatic complaints, cervical spondylosis, upper respiratory infections, and epilepsy	AH, Adhyayam 21

Continued

Table 2.3. Continued

SN	Formulation	Used in	Reference
14.	*Brahmarasayanam*	Immune deficiency, rejuvenative and restorative memory enhance	AH, Adhyayam 39
15.	*Brihacchagaladi Ghritam*	Aphrodisiac. Demyelinating disorders and general debility	BR, Adhyayam 74
16.	*Chandraprabhavatika*	Diabetes mellitus, genito-urinary tract infections, dysuria, renal calculi, prostatitis, benign prostatic hypertrophy (BPH) and emaciation	BR, Adhyayam 37
17.	*Cheriyachinchadi Leham*	Anaemia, loss of appetite and indigestion	Proprietary medicine
18.	*Cheriyanarayana Tailam*	Management of resultant from spasm, inflammation, obstruction and degeneration	BR, Adhyayam 26
19.	*Chitrakaleham*	Hemorrhoids, skin diseases, splenomegaly and indigestion	AH, Adhyayam 8
20.	*Dadimashtakachurnam*	Irritable bowel syndrome, malabsorption disorders, amoebic dysentery	AH, Adhyayam 9
21.	*Dantiharitakileham*	Rejuvenative and restorative in debilitating disorders	AH, Adhyayam 14
22.	*Dasamulaharitakileham*	Generalized oedema, genito-urinary diseases, splenomegaly, ascites and hypertension	AH, Adhyayam 17
23.	*Dasamularasayanam*	Acute breathlessness, productive cough	Proprietary medicine
24.	*Dasamularishtam*	Relieves fatigue instantly, restores pre-morbid health status, cardio pulmonary ailments, dyspepsia, loss of libido and quick acting herbal tonic during convalescence. Improves stamina and vitality. Useful in post-natal care	BR, Adhyayam 74
25.	*Dasanakantichurnam*	Maintain oral hygiene	Proprietary medicine
26.	*Dehaposhanayamakam*	Rejuvenative. Emaciation due to debilitating disorders	Proprietary medicine

Continued

Table 2.3. Continued

SN	Formulation	Used in	Reference
27.	*Devadarvyarishtam*	Diabetic neuropathy, urinary tract infection, eczema, indigestion and constipation	BR, Adhyayam 37
28.	*Dhanwantaram Gulika*	Asthma, dyspnoea, hiccup, cough and gastric discomfort	Sahasrayogam (SY)
29.	*Dhanwantaram Tailam*	Neuromuscular and skeletal diseases of vata origin, prevention of vatavyadhis, pre- and post-natal care. Dhanwantharam Mezhukupakam: For snehavasthi, matravasti in all vata-vitiated conditions	AH, Adhyayam 2
30.	*Dhanwantararishtam*	Vatavyadhis of varied aetiology especially of traumatic origin. Useful in post-natal care	AH, Adhyayam 2
31.	*Dhurdhurapatradi Keratailam*	Hairfall, dandruff, infantile eczema and psoriasis	AKD, Adhyayam 30
32.	*Draksharishtam*	Anaemia, breathlessness, fatigue, dyspepsia and constipation	SS, Adhyayam 10
33.	*Eladi Churnam*	Healthy skin care. Helps to restore complexion and radiance of skin useful in the treatment of allergic skin rashes, acne and dermatitis	AH, Adhyayam 15
34.	*Eladi Keratailam*	Chronic catarrh, kapha-vata predominant disorders and allergic pruritis. Enhances complexion	AH, Adhyayam 15
35.	*Elakanadi Kashayam*	Rajayakshma, diseases of respiratory system	SY, 137
36.	*Gandhakarajarasayanam*	Chronic skin diseases and skin manifestations in auto-immune conditions including scleroderma, rheumatoid arthritis, etc.	YR
37.	*Gopichandanadi Gulika*	Infantile eczema especially in children, fever, seizure disorder and upper respiratory infection	Proprietary medicine

Continued

Table 2.3. Continued

SN	Formulation	Used in	Reference
38.	*Gorochanadi Gulika*	Fever, vertigo, fatigue, respiratory distress, hiccup, convulsions and sannipathika (multi-system involvement) conditions of acute and chronic diseases.	Proprietary medicine
39.	*Himasagaratailam*	Encephalopathy, attention-deficit disorders, insomnia, headache, manias, depression and fatigue due to over-exertion	BR, Adhyayam 26
40.	*Jirakadyarishtam*	Relieves post-partum fatigue instantly and useful in dyspepsia	BR, Adhyayam 69
41.	*Kalyanakaghritam*	Psychic disorders, seizure disorder, anaemia, infertility, impairment cognitive, poor concentration and diabetes mellitus	AH, Adhyayam 6
42.	*Kanchanaraguggulu*	Thyroid malfunction, cystitis and anorectal disorders	AFI, Part I
43.	*Kumaryasavam*	Diabetes mellitus, anaemia, diseases of liver, dysfunctional uterine bleeding and menopausal syndrome	SS, Adhyayam 10
44.	*Kusmandarasayanam*	Allergic and persistent cough, respiratory ailments, deficient immune conditions, hoarseness of voice and emaciation in children	AH, Adhyayam 3
45.	*Lodhrasavam*	Management of diabetes, obesity, anaemia and related conditions	AH, Adhyayam 12
46.	*Madhookasavam*	Malabsorption syndrome and indigestion	AH, Adhyayam 10
47.	*Madhuyashtyadi Tailam*	Chronic rheumatism and burning paresthesia	AH, Adhyayam 22
48.	*Mahabala Tailam*	Male and female infertility, post-natal care, fracture and dislocations and seizure disorders	BR, Adhyayam 26

Continued

Table 2.3. Continued

SN	Formulation	Used in	Reference
49.	*Mahadhanwantaram Gulika*	Ante-natal care, acid peptic disease and gastric discomfort	Proprietary medicine
50.	*Mahakalyanaka Ghritam*	Psychic disorders, seizure disorder, anaemia, infertility, lack of intelligence, poor concentration, diabetes mellitus and toxicity in the body	AH, Adhyayam 6
51.	*Maharajaprasarani Tailam*	Neurological and rheumatic complaints, cervical and lumbar spondylosis	BR, Adhyayam 26
52.	*Mahavilwadi Leham*	Anorexia, dyspepsia and GERD	Proprietary medicine
53.	*Manjishthadi Lailam*	Headache, insomnia and ophthalmic diseases	SY
54.	*Mridwikarishtam*	General debility, emaciation, anaemia, worm infestations, diseases of head and neck	SS, Adhyayam 10
55.	*Mulakadi Kashayam*	Allergic and infective dermatitis in children	Proprietary medicine
56.	*Mulakadyarishtam*	Cellulitis and dermatitis of varied aetiology	Proprietary medicine
57.	*Mustadi Marmakashayam*	Fracture and dislocations, traumatic diseases, muscular sprain and osteoporosis	Proprietary medicine
58.	*Narayanagulam*	Constipation associated with diabetes mellitus and dysuria	Proprietary medicine
59.	*Nasikachurnam*	Anosmia, congestive nasal conditions	Proprietary medicine
60.	*Nisosiradi Tailam*	Diabetes and diabetic carbuncles	Proprietary medicine
61.	*Putikaranjasavam*	Hemorrhoids, indigestion and anaemia	AH, Adhyayam 8
62.	*Rohitakarishtam*	Splenomegaly, hepatomegaly and dermatosis	BR, Adhyayam 41
63.	*Saraswatarishtam*	Enhances cognitive skills, helps management of dementia and psychomotor retardation	BR, Adhyayam 73

Continued

Table 2.3. Continued

SN	Formulation	Used in	Reference
64.	*Saribadyasavam*	Diabetes mellitus, carbuncles, rheumatism, infections and genito-urinary diseases	BR, Adhyayam 38
65.	*Satavarigulam*	Leucorrhea, menorrhagia, dysuria, general debility in women and hypogalactia	Proprietary medicine
66.	*Sitopaladi Churnam*	Tastelessness, numbness of tongue, indigestion, fever, cough and burning sensation in palms and soles	SS, Adhyayam 6
67.	*Stanyajananarasayanam*	Hypogalactia	Proprietary medicine
68.	*Sukhabhedi Gulika*	Laxative	Proprietary medicine
69.	*Swasaharaleham*	Dyspnoea and other respiratory ailments	Proprietary medicine
70.	*Talisapatradi Leham*	Indigestion, respiratory ailments, cough, breathlessness and hoarseness of voice	AH, Adhyayam 10
71.	*Talisapatradivatakam*	Indigestion, respiratory ailments, cough, breathlessness and hoarseness of voice	AH, Adhyayam 10
72.	*Triphaladi Tailam*	Headache, rhinitis, baldness, premature graying and splitting of hairs	Proprietary medicine
73.	*Trivril Leham*	Constipation	AH, Adhyayam 2
74.	*Valiya Amritadi Tailam*	Chronic sinusitis, abscess and carbuncles	Proprietary medicine
75.	*Valiya Arimedastailam*	Tooth ache, painful infection of teeth and gums and loose tooth – can be used for gandusham	AH, Adhyayam22
76.	*Valiya Madhusnuhirasayanam*	Chronic skin diseases including eczema, atopic dermatitis, psoriasis, purulent and persistent infections and non-healing ulcers	SY

Continued

Table 2.3. Continued

SN	Formulation	Used in	Reference
77.	*Valiya Narayana Tailam*	Vatavyadhi of multiple causes and symptoms, deafness, male sub fertility and fractures	BR, Adhyayam 26
78.	*Valiya Sahacharadi Tailam*	Parkinsonism, muscle wastage, foot drop, sciatica, lumbar spondylosis and burning paresthesia	AH, Adhyayam 21
79.	*Varavisaladi Kashayam*	Psychic disorders, seizure disorder, anaemia, infertility, lack of intelligence, poor concentration, diabetes mellitus and toxicity in the body	AH, Adhyayam 6
80.	*Vasishtharasayanam*	General debility, respiratory ailments including bronchial asthma, hiccup and cough	AH, Adhyayam 3
81.	*Vayugulika*	Asthma, cough, hiccup, spasm, dyspnoea, convulsions and epilepsy	Proprietary medicine
82.	*Vidaryadi Leham*	Helps to build-up tissues, management of cardio pulmonary diseases, muscle cramps, post-natal care and leucorrhoea	Proprietary medicine
83.	*Vilwampachottyadi Tailam*	Improves vision, gives coolness and comfort to the head and prevents cold	SY
84.	*Viswamritham*	Post-natal care. Fever, lack of hunger/appetite and loose bowels	Proprietary medicine
85.	*Vyoshadi Vatakam*	Cough, rhinitis, hoarseness of voice and related allergic manifestations	AH, Adhyayam 20

AFI – Ayurvedic Formulary of India (Anonymous, 2003); AH – Ashtangahridayam (Tripathi, 2009); AKD – Arogya Kalpa Drumam (Varrier K.R, 2005); BR – Bhaishajya Ratnavali (Mishra, 2011); SS – Susruta Samhita (Yadavajitrikamji, 2008); SY – Sahasrayogam (Nishteswar and Vidyanath, 2023); YR – Yogaratnakaram (Shashtri, 2010) plays a role in *Ajeya Ghrita*, aiding in the management of poisonings, while in *Tarkshya Agada*, it is applied for the treatment of snake bites. Moreover, cardamom is part of *Bhadrodaya Anjana*, an ayurvedic eye ointment, and is included in *Kalyanaka Ghrita*, which addresses infertility and paediatric idiopathic syndromes. Lastly, cardamom is an ingredient in *Eladi Rasayana*, used in the treatment of tuberculosis. Additionally, cardamom is a component of *Phala Ghrita*, employed in addressing *Grahadosha*, a paediatric idiopathic syndrome, and it is also beneficial in treating female infertility.

According to Vagbhata, author of *Ashtanga Hridaya*, the third most important text of Ayurveda, cardamom is an ingredient in various ayurvedic formulations used to treat a range of conditions (Tripathi, 2009). For bronchial asthma (*Kasa*), hiccups (*Hikka*) and other respiratory disorders (*Shwasa*), cardamom is included in remedies such as *Amritaprasha Ghrita* and *Kasanashaka Avaleha*. In the context of *Kaphaja Madatyaya*, a form of alcohol poisoning, cardamom is part of *Ashtanga Lavana*. It is also a component of *Patoladi Churna*, which is used to manage irritable bowel syndrome (*Grahani*), and of *Hingvadi Ghrita* for treating abdominal tumours associated with *Vata Dosha* (*Vataja Gulma*). Additionally, Vagbhata highlights cardamom's role as an *Anupana Dravya*, a vehicle for delivering therapeutic substances in *Virechana* yoga, an emetic preparation. Furthermore, cardamom is utilized in *Koshatakyadi Niruha Vasti*, a treatment targeting diseases caused by an imbalance of the *Kapha Dosha* (*Kaphaja Roga*). Cardamom is an ingredient found in *Bhutrarao Ghrita*, which is used to treat paediatric idiopathic syndrome known as *Grihabadha*. Additionally, cardamom is included in *Bhaskara Anjana*, a formulation utilized for addressing visual impairments such as *Timira* (vision disturbances), *Kacha* (cataract) and *Arma* (pterygium). Furthermore, cardamom is recognized as a *Prakshepa Dravya*, or additive component, in *Vajikarana Yoga*, which refers to aphrodisiac preparations (Tripathi, 2009).

2.7 Conclusion

The fruits (capsules) of *E. cardamomum* (L.) Maton, belonging to the Zingiberaceae family, have historically been utilized in traditional medicine to address a range of health issues, including asthma, infections of the teeth and gums, cataracts, nausea, diarrohea and various cardiac, digestive and kidney disorders. The versatility of cardamom capsules offers numerous health benefits, integrating the wisdom of traditional medicinal practices with contemporary pharmaceutical approaches. In the Ayurvedic system of medicine, various health conditions are effectively treated using specific remedies. Notable applications include treating infections of the teeth and gums, addressing throat ailments, alleviating lung congestion, managing pulmonary tuberculosis and reducing inflammation of the eyelids. Additionally, Ayurveda offers solutions for gastrointestinal disorders, promoting kidney health and addressing issues related to impotency, erectile dysfunction and premature ejaculation. It is also used in the management of gallbladder stones among other conditions.

Genetic Resources and Crop Improvement of Cardamom

3

H.J. Akshitha*, M.S. Shivakumar and Honnappa Asangi

ICAR-Indian Institute of Spices Research, Regional Station, Appangala, Madikeri, Kodagu – 571 203, Karnataka, India

3.1 Genetic Resources

3.1.1 Centre of origin and diversity

Cardamom probably originates from the evergreen rainforests of the Western Ghats in South India, particularly in regions like the Silent Valley and the Cardamom Hills, located in present-day Idukki district of Kerala. These areas still harbour wild populations of cardamom, supporting its origin theory (Ravindran *et al.*, 2002). Cardamom was traded as a forest product from these hills during the early spice trade from the West Coast of India. Commercial cultivation began in the 18th century, largely driven by the rising demand and British colonial interests. By the mid-1700s, the Cardamom Hills came under the control of the Travancore kingdom and plantation development accelerated. Early plantations were established around 1800 by Tamil landlords, spreading into nearby regions like Bodinayakanur and Theni in Tamil Nadu (Sivanandan *et al.*, 1986). The British dominated tea and coffee plantations, while cardamom cultivation remained in Tamilian ownership. Cardamom thrived in the lower, shaded forest areas while tea cultivation was widespread at higher altitudes. During the 1880s, fluctuations in cardamom prices forced the Travancore state to abandon its monopoly over cardamom trade, replacing it with a land tax system. Alleppey emerged as a major trading port, giving rise to the name 'Alleppey green cardamom'. Over time, export dominance shifted to other ports like Bombay (Murugan *et al.*, 2021). Later, cardamom cultivation spread to Karnataka's hilly districts such as Coorg, Chikkamagalur and Hassan, where it was introduced as an intercrop in coffee plantations.

*Corresponding author: akshi.hosahalli10@gmail.com

Cardamom cultivars and varieties are categorised based on panicle growth habit as Malabar (prostrate), Mysore (erect) and *Vazhukka* (semi-erect) types. The germplasm collections of cardamom harbour a diverse group of collections with variation in plant height, panicle growth habit, tiller production, panicle branching, capsule shape, size, colour and quality traits (Table 3.1).

Table 3.1. Distinguishing characteristics of the three major cultivar groups of cardamom.

Character	var. Malabar	var. Mysore	var. Vazhukka
Adaptability	Lower elevations (600–900 m MSL)	Higher elevations (900–1200 m MSL)	Widely adaptable across elevation ranges
Primary cultivation areas	Karnataka	Kerala and parts of Tamil Nadu	Predominantly Kerala
Plant growth habit	Medium, moderately vigorous	Tall and robust	Robust, intermediate between Malabar and Mysore
Panicle orientation	Prostrate	Erect	Semi-erect
Capsule shape	Round to oblong	Bold, elongated	Round to oblong (intermediate type)
Leaf petiole length	Short	Long	Long
Capsule colour at maturity	Pale/golden yellow	Deep green	Green

(Source: Korikanthimath *et al.*, 2024)

Fig. 3.1. Three distinct cultivar groups in cardamom (a) Malabar, (b) Mysore and (c) Vazhukka.

3.1.2 Related species

Cardamom is a perennial, herbaceous monocotyledon belonging to the family Zingiberaceae. The genus *Elettaria* comprises 11 species, but only *Elettaria cardamomum* is found and cultivated in India. Another species, *Elettaria ensal* (Gaertn.) Abeyw., native to Sri Lanka, was previously classified as *E. cardamomum* var. major Thwaites. The remaining nine species occur in the wild, primarily in the Malaysian–Indonesian region (Anonymous, 2022). Closely related genera such as *Amomum* include *Amomum subulatum* (large cardamom), which is widely cultivated in the Eastern Himalayas. Wild *Elettaria* species from Southeast Asia, along with lesser-known taxa such as *Amomum pterocarpum*, *Amomum aculeatum*, *Hedychium coccineum*, *Hedychium flavescens* and species of *Alpinia*, are also part of this broader genetic pool. Recently, two new species of cardamom were identified from Kerala's Western Ghats. *Elettaria facifera* was identified from Periyar Tiger Reserve and *Elettaria tulipifera* was identified from the Agasthyamalai hills of Thiruvananthapuram and parts of Munnar.

3.1.3 Collections

Natural cardamom populations in the Western Ghats, including the Cardamom Hills, are declining due to forest ecosystem degradation and resulting changes in the evergreen rainforest climate. As a cross-pollinated crop primarily pollinated by honeybees, cardamom reproduces through both seeds and suckers. Despite this, ecosystem diversity in natural stands is now minimal (Ravindran and Madhusoodanan, 2002). Systematic collection and conservation of cardamom genetic resources began in the 1960s. Earlier, major explorations were carried out to collect wild genetic variability (Mayne, 1942) and to document the species' distribution and environmental influences (Abraham and Tulsidas, 1958). Although efforts have been made, conservation is challenging due to the spread of the *Katte* mosaic virus, other viral diseases, rhizome rot and blight.

Cardamom germplasm accessions were collected from the forest area/natural habitat, farmers' fields by the research organizations and are being conserved in the field gene banks. Indian Council of Agricultural Research (ICAR) - Indian Institute of Spices Reserach (IISR), Regional Station, Appangala being the National Active Germplasm Site (NAGS) for cardamom is conserving the germplasm collected from other research organizations along with its own collections. The collections of the research organizations must be submitted to the NAGS centre for obtaining the indigenous collection (IC) number from National Bureau of Plant Genetic Resources (NBPGR). The individual organization of germplasm holding in India is provided in Table 3.2.

3.1.4 Conservation and characterization

Systematic collection and conservation of cardamom genetic resources started in the 1960s. In the 1950s, two major surveys were conducted for collecting the genetic resources from wild habitat (Mayne, 1942) to understand

Table 3.2. Status of cardamom germplasm holdings in India.

Organization	Conservatory	Germplasm holdings
Indian Council of Agricultural Research	Indian Institute of Spices Research, Regional Station, Appangala, Kodagu, Karnataka	625
Spices Board	Indian Cardamom Research Institute (ICRI), Myladumpara, Idukki, Kerala	545
	ICRI Regional Station, Sakaleshapur, Karnataka	264
University of Agricultural and Horticultural Sciences (UAHS), Shivamogga	Zonal Agricultural and Horticultural Research Station, Mudigere, Karnataka	132
Kerala Agricultural University (KAU)	Cardamom Research Station, Pampadumpara, Idukki, Kerala	188

the geographical distribution and also the environmental impact on cardamom (Abraham and Tulsidas, 1958). Research organizations are working on improvement of cardamom by conserving the germplasm (Table 3.2). In general, germplasm is conserved by two ways, i.e. *in situ* and *ex situ*. In *in situ* conservation, cardamom accessions are conserved in their natural forest habitats with minimal or no human interference, as observed in areas like Silent Valley in Kerala and Malai cultivation regions of Bhagamandala, Karnataka. On the other hand, *ex situ* conservation involved maintaining germplasm accessions through the following approaches: Short-term conservation is achieved via *in vitro* storage under slow-growth conditions using micropropagated plants (Nirmal Babu *et al.*, 1999a,b); long-term conservation involves cryopreservation in liquid nitrogen at −196°C (Chaudhary and Chandel, 1995). Additionally, field gene banks serve as another important method for conserving germplasm resources, *ex situ*.

The utility of germplasm material depends not only on its collection and conservation but also on the characterization of available genetic variability. A preliminary set of descriptors for cardamom germplasm evaluation was developed by Dandin *et al.* (1981), which laid the foundation for standardizing germplasm evaluation. This effort was later strengthened by the International Plant Genetic Resources Institute (IPGRI, Rome), which published a comprehensive descriptor for cardamom (IPGRI, 1994), enabling more systematic evaluation and documentation of genetic resources. In India, DUS guidelines published by the Protection of Plant Varieties and Farmers' Rights Authority (PPV&FRA) describe 26 quantitative and qualitative characters (PPV&FRA, 2007) for cardamom. These descriptors are being adopted or used by the researchers to characterize the cardamom germplasm. The genotypes that are registered under PPV&FRA are being maintained in the cardamom field gene bank of PPV&FRA at ICAR-IISR, Regional Station, Appangala.

Prasath and Venugopal (2004) evaluated 310 cardamom germplasm accessions for 16 characters and classified them into three distinct clusters, indicating strong ancestral similarities among established varieties. Among the Malabar types, the highest variability was recorded for yield per plant (coefficients of variation [CV] = 92.80%), followed by number of panicles per plant (78.76%) and capsule width (69.77%). In contrast, seeds per capsule and leaf length exhibited low variability, with CV values of 9.69% and 15.74%, respectively. In Mysore types, high variability was observed for yield, number of panicles per plant and internodal length, whereas leaf length and leaf width exhibited low variability. Whereas, high variability was recorded for characters such as yield, number of panicles per plant and tiller-bearing capacity per plant in *vazhukka* types. Subsequent studies by the same authors on 23 accessions with compound panicles revealed high CV for key yield traits, notably yield per plant (77.94%) and branches per panicle (49.40%), with the lowest CV recorded for dry recovery percentage (7.73%). Two accessions, IC 547214 and IC 349544, were identified as promising due to their desirable yield-contributing characters. Additionally, accession IC 349553, which exhibited extensive branching at the proximal end of the panicle, and IC 547180, which retained dark green capsules even after drying, were found to be unique (Prasath and Venugopal, 2009). Fourteen cardamom genotypes, including three popular farmers' selections, were evaluated at high altitudes in Karnataka from 2003 to 2007. The farmers' selections yielded on par with the released variety 'IISR Suvasini'. Dry yield over 3 years ranged from 524.35 kg ha^{-1} ('APG 387') to 1140.54 kg ha^{-1} ('APG 250'), with 'APG 250' and 'APG 416' consistently recording higher yields across all 3 years (Prasath *et al.*, 2009a). The core traits strongly associated with yield improvement include panicles per plant, fresh capsule weight per plant, nodes per panicle and internodal panicle length (Gopal *et al.*, 1990, 1992).

Extensive germplasm characterization across different research centres has revealed substantial variability in agronomic traits. Krishnamurthy *et al.* (1989) identified 26 promising genotypes at the Regional Research Station, Mudigere, while Sudharsan *et al.* (1988) documented 29 distinct types at the Indian Cardamom Research Institute, Myladumpara. Subsequent evaluations by Padmini *et al.* (1999), Prasath *et al.* (2001), Sharon *et al.* (2020) and Honnappa *et al.* (2025) confirmed wide variability in essential traits such as number of tillers and panicles, capsule number, panicle length and overall yield per plant. Raj *et al.* (2000) in an evaluation of 72 accessions reported that multiple branch panicle type (MBP) recorded the highest wet and dry yields of 1050.5 g and 243.1 g $plant^{-1}$, respectively, whereas Clone-57 registered the lowest wet (250.8 g $plant^{-1}$) and dry yield (159.2 g $plant^{-1}$). However, the yield consistency of multibranch panicle (MBP) and compound panicle (CP) was poor, as indicated by high coefficients of variation of 79.0% and 100.6%, respectively. In contrast, accession S-1 exhibited a high degree of yield stability over years with a lower CV of 37.0%. The higher dry yield in S-1 could be attributed to a greater number of seeds per capsule compared to MBP and CP, whereas the higher wet yield in MBP and CP was mainly due to the branching nature of panicles, resulting in an increased number of capsules per panicle. Prasath *et al.* (2001) evaluated 119 Malabar types of cardamom genotypes and it was observed that characters such as yield per plant and number of capsules per plant, which exhibited high genotypic coefficient of variation heritability and genetic advance, can be effectively utilised in selection. It has been suggested that characters with high heritability coupled with high genetic advance respond better to selection than those with high heritability but low genetic advance. Sharon *et al.* (2020) evaluated 117 field gene bank (FGB) accessions and observed substantial phenotypic variability. Among the genotypes, FGB 13 recorded the fresh capsule weight (399.61 g plant^{-1}) and number of capsules per plant (244.20), indicating their potential value in selection and breeding programmes. Honnappa *et al.* (2025) evaluated another set of 80 field gene bank accessions and observed significant variability among the accessions for key traits such as plant height, bearing tillers, panicle number and capsule yield, reflecting considerable genetic diversity in the germplasm. Accession FGB 203 was superior, recording the highest plant height, bearing tillers, panicles, capsule number and fresh capsule yield. Cluster analysis grouped the accessions into three distinct clusters, with Cluster II comprising accessions with the most desirable agronomic performance.

Evaluation of 70 small cardamom genotypes revealed wide morphological and yield-related variability, indicating rich genetic diversity. Plant height ranged from 110 to 310 cm, with IC 547186 being the tallest. IC 547214 recorded the highest number of bearing tillers, while IC 547205 was superior for panicle length, nodes per panicle, capsule number and fresh capsule yield per plant. Genotype IC 584093 was the next best for yield traits. Considerable variation was also observed in capsule characters, with IC 584096 producing larger capsules with more seeds, highlighting promising genotypes for yield improvement and breeding programmes. Similarly recorded a highly significant correlation between fresh capsule yield and capsules per plant (0.999) and panicle length (0.762), emphasizing their value as key selection criteria in cardamom breeding (Senthil Kumar *et al.*, 2018). Preethy *et al.* (2022) assessed phenotypic diversity in Western Ghats accessions, highlighting significant variability in yield and vegetative traits. Among the evaluated 67 small cardamom accessions, Veeraputhran recorded the highest plant height and the maximum number of panicles per plant (46.66). The highest variability was observed for number of capsules per plant (64.40%), followed by panicle length and number of panicles per plant. MBP produced the highest number of capsules per plant (3139.2), mainly due to its branched panicle nature. Moderate variability was recorded for number of internodes per panicle, number of panicles per tiller and number of tillers per plant, with Type 103 registering the highest number of internodes per panicle (38) and longest panicle length. In contrast, plant height and number of seeds per capsule exhibited the least variability, indicating comparatively stable traits across accessions. In cardamom, yield is directly influenced by key traits including the number of bearing tillers, number of panicles, panicle length, number of cincinni per panicle and the number of capsules per cincinni.

Cardamom is valued for its essential oil content, which generally ranges from 6.5% to 10.5%. Its characteristic sweet aroma is mainly attributed to α-terpinyl acetate, linalyl acetate and linalool, while the medicinal properties are largely due to 1,8-cineole. Chemical profiling has shown that the Mysore type is richer in α-terpinyl acetate, giving it a sweeter flavour, whereas the Malabar type contains higher levels of 1,8-cineole, resulting in a sharper, camphor-like aroma (Sarathkumaran *et al.*, 1985). Germplasm evaluations have identified several promising high essential oil–yielding accessions, including Acc. 221 and Acc. 218, which consistently produced 7.8% oil over 3 years (Prasath and Venugopal, 2004) and PS-5 (essential oil 8.4%, oleoresin content 11.4%) and PS-12 (essential oil 8.4%, oleoresin 11.6%) (Raj *et al.*, 2000). Further screenings revealed Cl 671, Cl 779 and Cl 731 as high essential oil–yielding types (8.5%) (Korikanthimath *et al.*, 2000b). Ashokkumar *et al.* (2020) assessed 65 accessions for essential oil content and it ranged from 2.2% to 9.6%. Five accessions recorded the highest volatile oil content, namely HY-3 and HY-4 (9.6%), S1 × GG (Type 2) and SAM-5 (9.3%) and PPK-2 (9.1%). The higher essential oil content observed in these accessions may be attributed to a greater number of seeds per capsule, optimal capsule maturity and the curing period.

Essential oil yield and quality are strongly influenced by capsule maturity. Capsules harvested at the immature (95–109 days after flowering) or physiologically mature stages (110–124 days) produced 20–30% more essential oil and retained better green colour than those harvested fully ripe (125–140 days). Among the genotypes, IC 547185 recorded the highest essential oil yield at the immature stage. In IC 349646, IC 349650, IC 349396, IC 349589 and IC 547136, the essential oil yield at the immature and physiologically mature stages did not differ significantly. The highest α-terpinyl acetate content was recorded in IC-349646 and IC-349650. IC-349396 exhibited the lowest 1,8-cineole content, along with comparatively higher levels of linalool and nerol than the other genotypes (Korikanthimath, 1993; Leela *et al.*, 2008). Therefore, harvesting at the physiologically mature stage is recommended to maximize essential oil yield while maintaining superior capsule colour and quality.

Essential oil components of nine farmer-selected cardamom varieties *Njallani* Green Gold, Thiruthali, Panikulangara Green Bold No.1, Wonder Cardamom, Elarajan, Arjun, Pappalu, PNS Vaigai and Pachaikkai along with ICRI-5 as a control, were analysed to assess genotype quality. Essential oil content was significantly higher in *Njallani* Green Gold (9.18%), followed by Wonder Cardamom (8.36%). GC-MS (gas chromatography–mass spectrometry) profiling identified 25 compounds accounting for about 90–93% of the total essential oil. The highest proportion of 1,8-cineole was observed in PNS Vaigai (38.04%) and ICRI-5 (38.99%), whereas Panikulangara Green Bold No.1 recorded the lowest level (24.30%). In contrast, α-terpinyl acetate was highest in Panikulangara Green Bold No.1 (43.39%), followed by Thiruthali

(39.21%), and lowest in ICRI-5 (35.97%). Panikulangara Green Bold No.1 also exhibited higher proportions of α-terpineol, geranyl acetate, geraniol and linalyl acetate compared to other genotypes. Sabinene content was highest in *Njallani* Green Gold (7.56%) and lowest in Elarajan (5.46%) (Paul *et al.*, 2024).

A study from Guatemala's most productive region evaluating 288 cardamom accessions revealed high morphological diversity (maximum Shannon diversity index, H′ = 2.02). Among the 17 morphological traits studied, key characteristics contributing to this diversity include plant height, capsule colour, number and length of panicles, rhizome colour and capsule-related traits. A comprehensive evaluation resulted in the selection of 14 superior accessions, which fell into three of six clusters formed through hierarchical analysis. To ensure representation of both major cardamom varieties ('Malabar', 'Mysore', '*Vazhukka*') and all cluster groups, 12 additional accessions were chosen. Altogether, 26 accessions were selected, combining high performance with broad genetic representation (Herrera-Gonzalez *et al.*, 2024).

3.1.5 Screening germplasm for biotic and abiotic stress

The germplasm accessions were also evaluated for specific traits apart from yield such as tolerance to biotic and abiotic stresses and genetic stocks were identified for climate resilience.

A survey in Kodagu, Karnataka, identified a rhizome rot–tolerant cardamom line (RR-1), which showed low disease incidence and high yield. It was later released as the variety IISR Avinash, noted for its disease resistance, high yield and dark green capsules (Venugopal *et al.*, 2006). Praveena *et al.* (2013) in a screening of 50 cardamom accessions against leaf blight disease identified six accessions, *viz.* IC-547222, IC-547223, IC-349645, IC-349649, IC-547158 and IC-349637 moderately resistant to leaf blight under field conditions and highly resistant to rhizome rot. Senthil Kumar *et al.* (2018) identified cardamom genotypes IC 547188, IC 547224, IC 584083, IC 547226, IC 584087, IC 547203, IC 584081, IC 547185, IC 584084, IC 547183, IC 547210, IC 547202 and IC 547204 that exhibited dual resistance to rhizome rot and leaf blight diseases. Screening of 180 cardamom germplasm accessions lead to the identification of eight accessions IC 349606, IC 349370, IC 349363, IC 349362, IC 547167, IC 349455, IC 349588 and IC 349364 as tolerant to cardamom thrips, with average capsule damage ranging from 13.7% to 16.7% (Jacob *et al.*, 2020).

Screening of cardamom accessions at ICAR-IISR, Regional Station, Appangala revealed variation in resistance to major diseases. Among 117 field gene bank accessions, FGB 35 was found resistant to leaf blight, while FGB 15 was highly resistant to rhizome rot (Sharon *et al.*, 2020). Evaluation of 80 accessions revealed that 27 genotypes were found to be resistant to leaf blight, whereas all accessions were moderately susceptible to susceptible for rhizome rot (Honnappa *et al.*, 2025). Venugopal (1999) identified virus-resistant cardamom plants from South Indian hotspots, confirming their resistance through multi-location screening. One superior, high-yielding line was later released as the variety IISR Vijetha.

Fifteen cardamom accessions were screened for resistance to major diseases and responses were correlated with the biochemical defence parameters. Among the accessions, Hybrid-6, Hybrid-17 and Hybrid-1 showed resistance to diseases and these could be utilized in crop improvement programmes. Among the biochemical defense parameters evaluated, peroxidase (PO) and β-1, 3-glucanase emerged as significant indicators of resistance to *Phytophthora* rot. In contrast, polyphenol oxidase (PPO) and chitinase played a major role in conferring resistance to clump rot and leaf blight. Resistance to *Fusarium* rot was primarily associated with elevated PO activity, while phenylalanine ammonia-lyase (PAL) and chitinase levels showed a strong influence and significant correlation with resistance to Phoma disease. Collectively, these biochemical markers can be effectively exploited for the identification of disease-resistant cardamom accessions and to support rapid and targeted breeding programmes in the future (Remya *et al.*, 2025).

Moisture stress during the dry months (December–May) adversely affects cardamom growth and yield (Hegde and Korikanthimath, 1996). Dry spells or moisture stress during these periods can lead to drying and eventual death of cardamom plants. Moisture stress also affects tiller emergence, inflorescence development and when inflorescences have already emerged, the lack of rain or irrigation severely reduces capsule set. In an evaluation of 12 cardamom clones under different shade and moisture regimes, clones P6 and P3 performed superiorly in yield (Gurumurthy *et al.*, 1996). Ankegowda and Krishnamurthy (2008) evaluated six high-yielding cardamom genotypes for moisture stress tolerance. Among them, Hybrid 36, Mysore 2 and Green Gold exhibited better adaptability under stress conditions, attributed to their greater leaf area, higher tiller number and increased dry weight. A study on eight elite cardamom lines showed that moisture stress reduced photosynthesis and yield, while antioxidant enzymes helped maintain cell water and gas exchange. IC 584058 exhibited bold capsules, minimal yield reduction and stress-adaptive essential oil changes (Ankegowda *et al.*, 2022a).

3.2 Floral Biology and Pollination

3.2.1 Floral morphology, structure and phenology

In cardamom, inflorescences emerge from the rhizome and consist of racemose panicles borne on cane-like peduncles, which are differentiated into nodes and internodes, each node bearing a cluster of flowers (cincinnus) (Madhusoodanan *et al.*, 2002). The panicle is a compound inflorescence of multiple cincinni, with bracts that are lanceolate and often fimbriate with age (Fig. 3.2a). Flowers are bisexual and zygomorphic, typically white with a violet-streaked labellum. The calyx is tubular (up to 2 cm long) and obscurely three-lobed, while the corolla has three lobes (1–1.5 cm long) with the dorsal lobe widest (Priyadarshan *et al.*, 1992). The labellum is obovate

Fig. 3.2. (a) Cardamom panicle; (b) honey bee pollinating cardamom flower; (c) cardamom capsules at different stages of maturity.

and three-lobed, serving as a landing platform for pollinators. The stamen is sessile with parallel thecae and a short connective crest; the ovary is inferior, trilocular, many-ovuled and glabrous, with axial placentation (Parameswar & Venugopal, 1974). Pollen grains are large (75–120 µm in diameter), spherical and sticky, facilitating insect-mediated transfer (Sinu and Shivanna, 2007). The stigma is funnel-shaped, encircled by cilia that aid in pollen capture and retention.

Cardamom exhibits a prolonged and non-synchronous flowering pattern. The phenological growth stages of inflorescence and flowering have been described in detail by Akshitha *et al.* (2024). Inflorescence emergence begins from the base of the tiller following the pre-monsoon showers (February–March), with peak panicle initiation observed after the onset of monsoon (June–August). Panicle development takes about 6–7 months. Flowering coincides with stages 504–509, continuing for 4–5 months, with the first flower opening after 60–80 days of panicle emergence. Peak flowering occurs from June to August, overlapping with capsule development and ripening (stages 710–819). Flowers open between 4.30 and 5.30 a.m., with anther dehiscence between 6.30 and 7.30 a.m. (Kuriakose *et al.*, 2009). Each flower withers within a day.

3.2.2 Floral biology, pollination and fruit set

Cardamom flowers are bisexual and generally self-compatible, yet cross-pollination predominates due to floral morphology and pollinator behaviour (Parameswar, 1973). The stigma's position relative to the anther and the shape of the labellum promote insect visitation and outcrossing. Although autogamy (self-pollination) is possible, effective fruit set is significantly higher under cross-pollination, which is mediated mainly by bees (Belavadi and Parvathi, 1998). Stigma receptivity and pollen viability are highest in the morning hours (8–10 a.m.), coinciding with peak pollinator activity. Pollination during this period yields up to 72% fruit set (Pattanashetti and Prasad, 1972).

Cardamom is entomophilous, and honeybees are the primary pollinators (Fig. 3.2b). The major pollinators include *Apis cerana*, *A. dorsata* and *A. indica* (Madhusoodanan *et al.*, 2002). The efficiency of pollination correlates with bee activity, which is influenced by humidity and floral density. Optimal fruit set occurs during humid months (June–September) and declines sharply in dry periods (December–March). A comparative study undertaken between wild and cultivated cardamom reveals a shift in pollinator assemblage due to domestication. It is found that wild cardamom relies primarily on solitary bees such as *Megachile* sp. and *Amegilla* spp., whereas cultivated cardamom is pollinated predominantly by social bees (*Apis cerana*, *A. dorsata*) (Sinu and Shivanna, 2007; Belavadi *et al.*, 1997).

Cardamom fruits are trilocular capsules that are oblong to globose and takes approximately 120 days after full bloom to reach full maturity (Krishnamurthy *et al.*, 1989) (Fig. 3.2c). Each capsule contains 12–32 seeds, enclosed in a mucilaginous aril. Seed colour changes from white (immature) to brown and finally black when ripe (Murugan *et al.*, 2016). Capsule development peaks until August and gradually decreases thereafter. Fruit set and yield decline with plant age as reproductive vigour reduces. Cardamom plants generally commence bearing within two to 3 years of establishment, with suckers bearing earlier than seed-derived seedlings. Economic yield starts at 3rd year after planting and, under good management, plantations remain productive for 10–12 years.

3.3 Cytogenetics

Cytogenetic studies in *E. cardamomum* provide essential insights into its genetic constitution, evolution and variability, which are critical for crop improvement. The first cytological investigation by Gregory (1936) reported a somatic chromosome number of $2n = 48$, which was later confirmed by Chakravarti (1948), Sharma and Bhattacharyya (1959). Cardamom is considered a balanced tetraploid, likely derived from a common ancestral basic number $x = 12$, shared with related genera such as *Globba*, *Phoemaria*, *Amomum* and *Alpinia* (Sudharsan, 1989). Chromosome number variation has been reported among various cultivated types: the Malabar variety has $2n = 48$, while the Mysore variety has $2n = 50$, suggesting that aneuploidy and chromosomal structural alterations contributed to varietal differentiation (Chandrasekhar and Sampath Kumar, 1986). Cytological and embryological evidence indicates an amphidiploid origin, likely from extinct wild progenitors (Pai, 1965; Panchaksharappa, 1966). This tetraploid structure has maintained meiotic stability, enabling consistent fertility and reproductive success. The amphidiploid origin, combined with structural uniformity, suggests that cardamom underwent polyploidization followed by diploidization to stabilize its genome. Recent studies using flow cytometry and chromosomal counts confirmed the tetraploid nature ($2n = 48$) across 30 accessions, including cv. Mysore, cv. Malabar and cv. *Vazhukka*, with no variation detected (Anjali *et al.*, 2016a). Nuclear DNA content (2C value) averaged 2.84 pg, corresponding to an estimated genome size of ~2783 Mbp, consistent with other Zingiberaceae species (Anjali *et al.*, 2016b).

3.4 Breeding

3.4.1 Breeding objectives

In India, several efforts are undertaken by different research organization for breeding cardamom to fulfil the following objectives:

- Development of high-yielding and stable genotypes with sustained productivity under diverse environments.
- Evolution of synchronised flowering types to minimize harvesting cycles and reduce labour requirements.
- Selection of plants producing bold, green capsules with a higher number of seeds per capsule.
- Incorporation of resistance/tolerance to major biotic stresses such as *Katte* (mosaic), rhizome rot and thrips.
- Development of tolerance to abiotic stresses, particularly drought.
- Improvement of quality traits, focusing on higher essential oil and oleoresin content, especially elevated levels of α-terpinyl acetate, which imparts the characteristic aroma and flavour of cardamom.

3.4.2 Breeding methods

Cardamom plants can be propagated both by sexual as well as asexual methods offering numerous opportunities to improve cardamom yields. Various crop improvement approaches used in cardamom are selection, heterosis breeding, mutation breeding and polyploidy breeding.

3.4.2.1 Selection

Selection is the most widely adopted method for genetic improvement in cardamom, as natural variability for yield and quality exists in the natural population due to cross pollination. Several improved and farmer-preferred varieties have been developed through this approach (Venugopal, 1999). Korikanthimath *et al.* (1999, 2000, 2002) evaluated seedling progenies and elite genotypes, identifying high-yielding types such as Entry 186, AS 6/3, Clone 779 and

S.1272–5. Their studies highlighted the importance of traits like compact panicles, number of capsules per panicle and capsule weight as key selection criteria (Korikanthimath and Mulge, 2005). Clonal selection has also led to the release of superior varieties such as IISR Avinash, noted for its high yield, dark green capsules and tolerance to rhizome rot (Venugopal *et al.*, 2006).

Extensive evaluations of promising genotypes and landraces have further strengthened selection programmes. Prasath *et al.* (2009b) identified APG 250 and APG 416 as high-yielding, while Hrideek *et al.* (2011) reported Ela Rani-2, Ela Rani-3 and Palakkudi as superior farmers' varieties. Comparative studies performed by Akhila *et al.* (2017) found Pappalu, Kalarikkal Bold and Pulari to outperform *Njallani* Green Gold and ICRI-2. More recent evaluations by Sadashiv Nadukeri *et al.* (2020) and Sharon *et al.* (2020) identified superior collections such as KMRD-2, CSS-800, RR-1, PDP-12 and FGB 13, the latter noted for its highest fresh capsule weight and capsule number per plant. These findings reinforce the scope of selection in developing high-yielding and quality genotypes in cardamom.

3.4.2.2 Heterosis breeding

Cardamom, being amenable to both sexual and asexual propagation, offers good scope for hybridization as a crop improvement strategy. The *Vazhukka* type is believed to be a natural hybrid between Malabar and Mysore types. The hybridization with species has shown variable compatibility depending on parental combinations (Venugopal and Padmini, 1999). The hybridization between varieties and cultivars have shown to combine desirable traits such as earliness, bold capsules, high yield, long panicles, resistance to leaf rot and multiple branching. Diallel crosses revealed that hybrid progenies were generally more vigorous than their parental lines (Krishnamurthy *et al.*, 1989). Heterosis studies in cardamom have demonstrated the potential of hybrid breeding. Padmini *et al.* (2000) reported significant positive heterobeltiosis and economic heterosis in traits like leaf length, breadth and plant height, with hybrids outperforming open-pollinated progenies and inbreds. Owing to cross-pollination nature of the crop, inbreeding depression was evidently observed in cardamom. Prasath *et al.* (2009) identified superior hybrids such as CCS 19 × NKE 19 and RR 19 × NKE 12, which showed higher yields and mosaic resistance than the check variety IISR Avinash. Senthil Kumar *et al.* (2017) assessed various hybrid combinations in cardamom and identified IC 547167 as moderately resistant to cardamom mosaic virus under artificial inoculation. This hybrid also exhibited high α-terpinyl acetate content, comparable to that of Appangala 1. In 2014, it was recommended for release as the variety Appangala 2, developed from the cross Appangala 1 × NKE 19. Akshitha *et al.* (2021) evaluated 23 inter-varietal F1 hybrids and reported nine superior hybrids, with ICRI 4 × IISR Vijetha performing best for both fresh (1090 g plant^{-1}) and dry yield (236 g plant^{-1}). These results highlight the scope of heterosis breeding in developing high-yielding, disease-resistant and quality cardamom hybrids.

3.4.2.3 Mutation breeding

Both physical mutagens such as X-rays and gamma rays, and chemical mutagens like maleic hydrazide and ethyl methane sulfonate, have been employed to induce mutations in cardamom. Selfed and open-pollinated progenies of M1 plants were screened against *katte* virus by repeated inoculations with its vector, leading to the identification of 12 tolerant plants (Bavappa, 1986). However, no stable mutant varieties have been reported in cardamom so far.

3.4.2.4 Polyplody breeding

Polyploidy was induced in cardamom by treating sprouting seeds with 0.5% aqueous colchicine solution, resulting in tetraploids ($2n = 4x = 96$). The induced polyploids exhibited distinct morphological changes such as gigantism, thicker cuticle, increased epidermal cell layers and enhanced wax deposition on leaves (Sudharshan, 1987; 1989).

3.5 Improved Varieties and Farmers Varieties

In addition to the varieties developed by research institutes and universities, farmers have also contributed significantly to cardamom improvement. They have developed and registered several distinctive varieties, including *Njallani* Green Gold, Thiruthali, Wonder Cardamom, Panikulangara Green Bold 1, Panikulangara Green Bold 2, White Flower, PNS Vaigai, Elarajan, Pappalu, Arjun, Pachaikkai and Thadathil Gold. These farmer-bred varieties reflect local selection for traits such as boldness of capsules, adaptation to specific microclimates and early maturity, complementing the efforts of formal breeding programs. Farmer's varieties are being registered by PPV&FRA, which will provide protection and right to the farmer to conserve, multiply and sell their variety (Table 3.3; Fig. 3.3).

3.6 Biotechnology in Cardamom

Conventional breeding in cardamom is constrained by its perennial nature, cross-pollination and long generation time; hence, biotechnological approaches help in understanding the cardamom genome to pivot the genetic improvement programme based on holistic genetic information, which could be utilised for realising better yield, quality and managing pests and diseases (Tyagi *et al.*, 2009).

3.6.1 Tissue culture and micropropagation

Micropropagation has been successfully standardized in cardamom for the large-scale production of disease-free and uniform planting materials.

Protocols using explants such as rhizome buds, immature inflorescences and shoot tips have been optimized for clonal propagation (Reghunath and Bajaj, 1992; Tyagi *et al.*, 2009). *In vitro* conservation protocols were developed for conservation of cardamom germplasm using half-strength MS medium without growth regulators but with addition of 15 g l^{-1} sucrose and mannitol each (Babu *et al.*, 1999). Direct plant regeneration was achieved on MS medium supplemented with 5 µM of 6-benzylaminopurine (BAP), producing up to 6.5 shoots per culture in 2 months and 15.1 shoots in 4 months, with simultaneous rooting, enabling the regeneration of about 30,375 plants per year from a single explant. *In vitro* conservation by the slow-growth method was accomplished on half-strength MS medium with 5 µM BAP, where nearly 70% of cultures survived for up to 18 months at 25 ± 2°C and showed successful regrowth upon reculturing. A high survival rate (96%) was recorded during hardening, and the regenerated plants grew normally under greenhouse conditions (Tyagi *et al.*, 2009). Somatic embryogenesis and plant regeneration from callus cultures have also been reported, offering possibilities for *in vitro* selection and genetic transformation (Alniemi, 2010). Shoot tips were used as explants in 45 media treatments and the best media for shoot proliferation was found to be MS medium supplemented with 4.4 µM of BAP and 2.32 µM of kinetin (Kn) with 5.83 shoots regenerated per explant. Rooting was successful on full strength MS basal medium and 80% survival percentage was achieved after acclimatization of well-developed plantlets (Vaidya *et al.*, 2020).

3.6.2 Molecular markers and genetic diversity

Microsatellite markers were developed for small cardamom (*E. cardamomum*) using selective hybridization enrichment. Out of 140 identified repeats, primers for 58 microsatellites were designed, with 44 successfully amplifying expected products. Six markers were polymorphic across 20 important genotypes, showing 2–7 alleles per locus (average 3.6) and polymorphic information content values of 0.14–0.38. Specific markers distinguished MCC7 (Pink tiller) and APG434 (MA18). Dendrogram and principal coordinate analyses revealed genetic similarities and identified PV1 and ICRI 1 as the most divergent genotypes. The markers showed 71.4% cross-generic transferability within Zingiberaceae (Cyriac *et al.*, 2016). Molecular marker techniques such as RAPD, ISSR, SSR, AFLP and EST-SSR have been utilised for assessing genetic diversity, varietal identification and germplasm characterisation (Anjali *et al.*, 2015; Preethy *et al.*, 2022). These studies revealed considerable genetic variability among cultivars and accessions collected from different geographical regions. The development of SSR and EST-SSR markers from transcriptomic and genomic data has enhanced the precision of genetic studies (Anjali *et al.*, 2015; Gaikwad *et al.*, 2023). Gaikwad *et al.* (2023) used 60 SSR markers for diversity analysis in 60 cardamom accessions. An average of 14.57 alleles per locus was detected, with allele numbers ranging from 4 to 30. Population structure analysis indicated a high degree of admixture, likely attributable to the predominantly cross-pollinated nature of cardamom. The identified SSR markers provide a valuable resource for developing gene- or trait-linked markers, facilitating marker-assisted breeding for crop improvement. The SSR loci information has been compiled into a publicly accessible database, 'cardamomSSRdb', for use by the broader cardamom research community.

3.7 Genomics and Transcriptomics

Recent advances in next-generation sequencing have provided valuable insights into the genetic architecture of cardamom. Gaikwad *et al.* (2023) reported whole genome sequencing of small cardamom variety *Njallani green gold*. The assembled genome length was 1.06 Gb (gigabases), which is close to the estimated genome size of cardamom and is close to *Musa* species. A total of 2, 50, 571 SSRs were identified of which 2, 18, 270 were perfect SSRs and 32, 301 were compound SSRs. In this study, a total of 38,400 genes were annotated in which 14, 731 genes were mapped to the Plant Transcription factor database.

Nadiya *et al.* (2017), through genome-wide transcriptomic analysis of cardamom, identified and reported several transcripts encoding 1,8-cineole synthase. The assembled transcripts were annotated against multiple databases (NCBI NR, Swiss-Prot, TrEMBL, Pfam, PlantCyc, KOG and TAIR10), revealing numerous unigenes corresponding to enzymes in key secondary metabolite pathways, including phenyl propanoid, isoprenoid, lignin and alkaloid biosynthesis. In addition to cineole synthase, the authors detected transcripts for limonene synthase and limonene-6-hydroxylase, enzymes responsible for limonene synthesis and its subsequent degradation, respectively.

In a follow-up study, the same research group (Nadiya *et al.*, 2018) compared gene expression between wild and cultivated cardamom genotypes. They observed that most genes in the terpenoid biosynthetic pathway, including cineole synthase, were downregulated in cultivated varieties compared to wild ones. Notably, limonene synthase expression was significantly higher in the wild genotype, a finding corroborated by volatile profiling, which showed limonene content to be approximately six times higher in wild cardamom essential oil than in cultivated types. Other monoterpene pathway genes identified included geranyl diphosphate diphosphatase (geranyl-diphosphate diphosphate lyase) and borneol dehydrogenase, along with several enzymes of the mevalonate pathway responsible for producing the precursors isopentenyl pyrophosphate (IPP) and dimethylallyl pyrophosphate (DMAPP). These discovered transcripts provide valuable full-length sequences that can be used to clone genes, develop transgenic cardamom plants and ultimately enhance the accumulation of important flavour compounds (especially 1,8-cineole and related monoterpenes) in cultivated seeds.

Table 3.3. List of improved varieties developed by research organizations and universities.

Variety/Hybrid	Type	Average yield (kg ha^{-1})	Salient features	Adaptation
ICAR-Indian Institute of Spices Research, Regional Station, Appangala, Madikeri, Karnataka				
Appangala 1	Malabar	745	Adaptable to intensive cultivation under mono and mixed cropping systems. Matures early, produces 89% of bold capsules	Suitable for cultivation in all cardamom growing tracts of Karnataka and Wayanad of Kerala
IISR Avinash	Malabar	848	Dark green capsules, tolerant to rhizome rot. Potential yield 1483 kg dry capsules ha^{-1}	Rhizome rot prone hotspots; Kerala & Karnataka
IISR Vijetha	Malabar	643	Resistant to *Cardamom mosaic virus* (katte disease)	Moderate rainfall and shade areas; Kodagu, Hassan, Chikkamagalur, North Wayanad
Appangala 2	Malabar	927	Hybrid, high yielding, and resistant to *cardamom mosaic virus/katte*	Suitable for cultivation in all cardamom growing tracts of Karnataka and Wayanad of Kerala
IISR Manushree	Malabar	550 (irrigated) and 360 (moisture stress conditions.)	Moisture stress tolerant cardamom variety developed through clonal selection. 50% of the capsules are >8 mm and is stable yielder	Cardamom growing tracts of Kerala and Karnataka
IISR Kaveri	Malabar	482 (irrigated) and 308 (moisture stress conditions)	Compact flowering cardamom genotype with 70% of the capsules are >8 mm and relatively tolerant to moisture stress	Cardamom growing tracts of Karnataka

Continued

Table 3.3. Continued

Variety/Hybrid	Type	Average yield (kg ha^{-1})	Salient features	Adaptation
Indian Cardamom Research Institute (ICRI), Myladumpara, Kerala				
ICRI 1	Malabar	656	Early maturing, bold capsules	South zone of Idukki, Kerala
ICRI 2	Mysore	766	Medium-long panicles, oblong bold capsules, tolerant to *azhukal*	Suitable for cultivation in Vandanmedu and Nelliyampathy of Kerala and Anamalai, Meghamalai of Tamil Nadu
ICRI 4	Malabar	961	Medium-sized panicles, globose bold capsules.	Lower Palani hills, Tamil Nadu; low rainfall areas
ICRI 5	Malabar	1543	High yield hybrid, moderately drought tolerant, 68% capsules are>7 mm	Kerala and parts of Tamil Nadu
ICRI 6	Malabar	1900	High percentage bold capsules	Kerala and parts of Tamil Nadu
ICRI 7	Vazhuka	1000	Deep green bold capsules, Moderately tolerant to *azhukal* disease	Kerala
ICRI Regional Station, Sakaleshapura, Karnataka				
ICRI 3	Malabar	599	Early maturing, pubescent leaves, oblong bold capsules, rhizome rot tolerant	Cardamom growing tracts of Karnataka
ICRI 8	Malabar	706	Ovoid bold pale green capsules	Karnataka
Zonal Agricultural and Horticultural Research Station, (University of Agricultural and Horticultural Sciences), Mudigere, Karnataka				
Mudigere 1	Malabar	275	Suitable for high density planting, capsules are pale green	Cardamom growing tracts of Karnataka
Mudigere 2	Malabar	475	Early maturing, round/oval bold capsules; high-density planting	Cardamom growing tracts of Karnataka

Continued

Table 3.3. Continued

Variety/Hybrid	Type	Average yield (kg ha^{-1})	Salient features	Adaptation
Mudigere 3	Malabar	900–1100	Oval-oblong capsules	Cardamom growing tracts of Karnataka
Cardamom Research Station (Kerala Agricultural University), Pampadumpara, Kerala				
PV 1	Malabar	500	Early maturing; short panicle, elongated slightly ribbed capsules	Kerala & parts of Tamil Nadu
PV 2	Vazhuka	982	Green bold capsules	Cardamom Hill Reserves of Kerala
PV 3	Malabar	1120	Oblong light green capsules	Cardamom Hill Reserves of Kerala

Fig. 3.3. Improved varieties of cardamom (a) Appangala 1; (b) IISR Avinash and (c) IISR Kaveri.

3.8 Future Prospects

The future of cardamom genetic resources and crop improvement lies in the effective conservation and utilization of its rich germplasm, including cv. Malabar, cv. Mysore and cv. *Vazhukka*, along with landraces and wild relatives. Emphasis will be on characterising and documenting traits such as disease resistance, drought tolerance and quality parameters. Advances in molecular breeding, including marker-assisted selection, genomic tools and gene-editing technologies, offer opportunities to accelerate the development of high-yielding, disease-resistant and quality varieties. Hybridization, heterosis breeding, polyploidy induction and mutation breeding will continue to create variability for yield, stress tolerance and desirable quality traits such as high α-terpinyl acetate content and bold green capsules. Future breeding efforts will also focus on developing climate-resilient and mechanisation-friendly genotypes suitable for high-density planting, organic cultivation and low-input systems. Integrating conventional breeding with precision agronomy, post-harvest management and global collaboration in germplasm exchange will further strengthen cardamom improvement programmes, ensuring sustained productivity, quality and adaptability under changing climatic and market conditions.

Advances in Cardamom Cultivation

4

G.A. Rajanna[1]*, S.J. Ankegowda[1], Nimisha Mathews[2] and H.C. Vikram[3]

[1]ICAR-Indian Institute of Spices Research, Regional Station, Appangala, Madikeri 571201, India
[2]Cardamom Research Station, Pampadumpara, Kerala Agricultural University, Idukki, India
[3]Pepper Research Station, Kerala Agricultural University, Panniyur, Kannur 670142, India

4.1 Traditional Cultivation Practices and Constraints

Traditionally, cardamom was cultivated in forest clearings and shaded plantations with minimal external inputs, relying on rainfed conditions, shade trees and organic matter recycling. While ecologically sound, this system produced modest yields, often below 250 kg ha^{-1}. In recent years, productivity has declined further due to soil fertility depletion, nutrient imbalances, pest and disease outbreaks (notably thrips, capsule rot and rhizome rot) and irregular rainfall patterns associated with climate change. Water stress during prolonged dry spells and inadequate irrigation infrastructures have aggravated yield instability (Ankegowda and Krishnamurthy, 2008). Additional challenges include degeneration of planting material, poor postharvest handling that lowers quality and limited access to technical knowledge. Socioeconomic constraints such as small landholdings, heavy labour dependence and market price volatility further undermine farm profitability. Overall, the traditional production system faces ecological and economic pressures that necessitate technological, agronomic and institutional interventions to sustain yields and enhance resilience in modern cardamom cultivation.

*Corresponding author: rajanna.ga6@gmail.com

4.2 The Need for Technological Advancements

The recognition that conventional production systems alone cannot meet the rising global demand for cardamom has driven intensive research towards improved cultivars, scientific crop management and postharvest innovations. In the past two decades, national institutions and state agricultural universities have made notable progress through coordinated efforts. Molecular characterization and DNA fingerprinting now guide breeding for yield, essential oil quality and disease resistance, while tissue culture supports the large-scale propagation of disease-free plants (Prasath *et al.*, 2010). Precision irrigation, fertigation and integrated nutrient and pest management have enhanced resource efficiency and ecological balance, complemented by improved post-harvest technologies such as low-temperature drying and advanced packaging materials that retain colour and valuable volatile compounds.

Given that cardamom thrives in sensitive highland ecosystems, changing climate patterns altered rainfall, temperature fluctuations and soil moisture deficits have led to yield decline and pest resurgence in traditional growing regions of the Western Ghats (Prasath *et al.*, 2009). Consequently, research has turned to climate-resilient systems built around micro-irrigation, organic mulching, shade regulation and varieties with improved resistance to pests (Korikanthimath and Mulge, 2005). Interdisciplinary studies integrating genomics, metabolomics, soil science and engineering are shaping future innovations. Notably, essential-oil profiling and Information and Communication Technology (ICT)-based extension tools are connecting science with growers, improving field-level decision making. Supported by initiatives such as the National Horticulture Mission and Mission for Integrated Development of Horticulture (MIDH), these advancements offer a framework for sustainable intensification. However, a comprehensive synthesis bridging genetic resources, propagation, crop management and marketing remain scarce. This review aims to integrate scattered knowledge and guide future strategies for modern, sustainable cardamom production systems.

4.3 Advances in Propagation and Nursery Management

The success in cardamom cultivation largely relies on healthy, high-yielding and genetically uniform planting material. As a highly heterozygous, cross-pollinated perennial, maintaining varietal purity through proper propagation is vital for sustained productivity. The propagation through seed often results in genetic segregation and decline of elite cultivars, whereas unscientific vegetative propagation methods heighten disease risks, including *Katte* virus and rhizome rot (Venugopal, 1999). Technological advancements in recent years have markedly improved propagation efficiency and plant health. Innovations such as standardised nursery protocols, clonal propagation frameworks, disease indexing and tissue culture-based micropropagation have strengthened the production of true-to-type and disease-free stock. It offers flexibility depending on cultivation scale, genetic uniformity needs and disease considerations. Collectively, these refined propagation technologies form the foundation of modern cardamom production systems and contribute significantly to improving genetic stability, productivity and sustainability across diverse growing regions.

4.3.1 Seed propagation: Challenges and improvements

Cardamom seeds, enclosed in a mucilaginous coat, exhibit poor germination due to dormancy and rapid viability loss. For germination, seeds from fully matured, bold capsules from healthy clumps are collected in September. Since seed viability declines rapidly, sowing within 15 days preferably in September (Karnataka) and November–January (Kerala and Tamil Nadu) ensures early and uniform germination. Acid scarification with 25% nitric acid improves germination percentage. Seeds are sown in beds and watered regularly and germination starts after 20–25 days. The seedlings at 3–4 leaf stage will be transplanted to the secondary nursery (IISR, 2022) for further growth. Innovations like bio-priming with *Trichoderma harzianum* or *Pseudomonas fluorescens* offer disease suppression, while solarisation of nursery beds effectively lowers pathogen load (Khulbe, 2019).

4.3.2 Vegetative propagation

Traditionally, farmers propagate cardamom through division of underground rhizomes, each containing one aerial shoot and a portion of the rhizome with roots. As cardamom is a cross-pollinated crop, using suckers for propagation is the preferred method to get true-to-type plants. About 2500–3000 suckers are required for planting one hectare. In clonal gardens, high-yielding lines are multiplied using planting units consisting of rhizome pieces with one old and one young sprout, which can be easily separated from established clumps. A spacing of 1.8 × 0.6 m is recommended, accommodating approximately 3800 clonal nursery plants per hectare (Nimisha *et al.*, 2021). Owing to its early bearing nature and 30–40% higher yield than seedlings, this method is becoming increasingly common among cardamom growers, as it not only ensures faster establishment but also risks transmitting soil-borne diseases if mother clumps are not properly selected. To ensure genetic purity and disease-free material, certified mother plants should be selected from uniform, high-yielding and true-to-type clones and fields should be free from viral and fungal diseases for at least 2 years.

Proper shade and irrigation are given, and each planting unit yields 32–42 suckers in 12 months (Crop production Guide Horticulture Crops, 2020). Before planting, suckers should be dipped in 0.3% copper oxychloride or 0.2% carbendazim for 30 min, followed by air drying to prevent disease incidence. In recent years, bioagents like *Trichoderma viride* (10 g/l suspension) have been used as eco-friendly alternatives. Chinnathambi *et al.* (2024) found that the application of Arbuscular Mycorrhizal (AM) inoculum at a rate of

10% to cardamom seedlings was most effective in enhancing seedling growth, nutrient absorption and disease resistance. The use of AM can aid in the production of healthy seedlings, enhance nutrient uptake and reduce reliance on chemical fertilizers and pesticides.

4.3.3 Micropropagation and tissue culture techniques

Owing to the limited multiplication rate in conventional methods and the spread of viral diseases, tissue culture has emerged as a sustainable solution for large-scale production of disease-free and genetically uniform planting material. Tissue-culture research in small cardamom has progressed through several standardized protocols developed by multiple researchers. Micropropagation techniques have been effectively employed for the rapid and large-scale multiplication of disease-free elite clones of cardamom (Reghunath and Bajaj 1992). There are various studies on the use of different explants such as rhizomes (Babu *et al.*, 1997), vegetative buds (Nadgauda *et al.*, 1983), somatic embryos (Manohari *et al.*, 2008), callus cultures (Rao *et al.*, 1982), immature floral buds (Kumar *et al.*, 1985), as well as shoot tips and nodal segments (Tyagi *et al.*, 2009). The addition of 4.0 mg l^{-1} NAA (Naphthalene Acetic Acid) and 1.0 mg l^{-1} BAP (6-Benzylaminopurine) to the growth medium led to a stimulating impact on callus formation (Reghunath and Bajaj, 1992; Rajkumar *et al.*, 2007).

Tissue culture propagation produces ready to plant material in 8–12 months (Babu *et al.*, 1999). To produce adequate numbers of homogeneous and vigorous seedlings, repetitive subculturing is essential. For *in vitro*-raised seedlings to become acclimated to their adjacent environments, hardening is essential (Malhotra *et al.*, 2021). Initially, rooted seedlings are placed in polybags that are filled in a 1:1:1 ratio with sterile growing media, such as cocopeat, sand and soil. For about 30 to 45 days, primary hardening is done in a greenhouse or mist chamber with 80 to 90% relative humidity and diffused light. Secondary hardening is done in a shade-net house with steady exposure to ambient environments for 30 to 45 days, with a total hardening period of 60 to 90 days (Reghunath and Bajaj, 1992). During the first establishing stage, 50–60% shade should be maintained, and deep planting should be avoided by keeping the collar region at soil surface (Rajkumar *et al.*, 2007). During the early stages of growth, shading is ensured from direct sunshine, strong winds, pests and diseases is vital, and proper mulching helps preserve soil moisture. Better survival, early establishment and increased yield of cardamom plants generated from tissue culture are ensured by repetitive monitoring and best nutrient management techniques.

4.3.4 Nursery nutrient and water management

Better nursery management is critical for promoting plant vigour and improving transplanting performance. According to Padmini *et al.* (2000b), effective approaches include adding farmyard manure (5 kg m^{-2}) together with a N:P:K solution at 100:50:50 mg/l by irrigation every 15 days. Application of biofertilisers like Azospirillum, Azotobacter and phosphate-solubilising bacteria (PSB) in potting mixtures significantly improved root initiation and enhanced chlorophyll content. Application of neem cake at 150–200 g per plant reduced soil-borne diseases.

After acclimatisation, tissue-culture plants should be transferred to polybags (20 × 25 cm) containing a sterilised soil:Farm Yard Manure (FYM):sand combination in a 2:1:1 ratio and kept in 50% shade for 45–60 days. Low light intensity during this period increases photosynthetic efficiency and plant tolerance. Planting the acclimatised plants in the main field is recommended at the early May–June (monsoon months), where partially shaded conditions are available. For uniform growth, 2 × 2 m spacing is advisable for tissue-culture plants.

Improved water management employing drip or mist-irrigation significantly enhanced root growth with relative humidity of 70–80% and soil moisture. To reduce transplant shock, shade should be carefully controlled and during hardening stage, plants should be progressively exposed to sunlight. During crop establishment, practices like weed/residue mulching, drip-irrigation and the application of biofertilisers significantly increase plant survival rates of more than 90%.

4.4 Advances in Crop and Nutrient Management Practices

Cardamom being a perennial crop of the humid tropics, demands precise management of soil, water, nutrients and shade to sustain high productivity. Traditional systems relied on organic inputs and empirical practices, but intensification and climatic variability have driven innovations in irrigation, fertigation, soil conservation and nutrient scheduling. Deep, well-drained loamy soils with high organic matter are ideal for cultivation of cardamom. The forest loam soil with good moisture-holding capacity is ideal for cardamom cultivation. Soils with poor drainage or a strong clay texture are inappropriate for cardamom since it cannot withstand waterlogging. The ideal pH range for soil is between 5.5 and 6.5, and slightly acidic soils promote better root development and nutrient availability. Cardamom thrives on steep, rolling terrain with elevations between 600 and 1500 m above mean sea level (Ankegowda *et al.*, 2022). Because they promote natural drainage and avoid water stagnation around the root zone, gentle-to-moderate slopes are ideal. To reduce soil erosion, terracing or contour bunds should be installed on steep slopes.

Cardamom seedlings are sown at 2 × 1 m on hill slopes along the contour and 2.1 × 2.1 m in flat fields, mild slopes and valley bottoms. In high-production technology demonstration plots, cardamom yield of more than 500 kg/ha in just 2 years was observed. Therefore, the variety and duration of the crop on the field should be taken into consideration while determining spacing. Malabar types are relatively smaller plants that require closer spacing, whereas Mysore and Vazhukka types are tall with large rhizome characteristics that require broader spacing. As the rate of fertiliser administration rose, the number of leaves and panicles grew; however, as plant density increased, the number of leaves and panicles decreased. The best growth and production (543.9 kg ha^{-1}) were obtained using the trench method of planting (60 × 30 cm) and a spacing of 2 × 1 m. When planted in places with high rainfall or irrigation, cv. Mysore and cv. *Vazhukka* often have a plant-to-plant distance of 3 × 3 m or 2.4 × 2.4 m. In Karnataka, a distance of 1.8 × 1.8 m or 2 × 2 m is appropriate for cv. Malabar (Korikathimath and Venugopal, 1989).

Recent studies emphasize effective conservation measures such as contour bunding, vegetative barriers using vetiver or citronella and mulching with forest litter or dry leaves. A 10 cm mulch layer has been shown to conserve moisture, moderate soil temperature and enhance yields by 15–25%. Annual incorporation of organic manures like farmyard manure (10 t ha^{-1}), composted coir pith (5 t ha^{-1}) or vermicompost (2 t ha^{-1}) improves soil cation exchange and microbial activity (Korikanthimath *et al.*, 2024). Green manure crops like *Crotalaria juncea* and *Sesbania aculeata* further enrich soil nitrogen and physical properties. Use of selective herbicide particularly oxyfluorfen (0.5 kg ha^{-1}) aids weed management when applied cautiously. Cardamom is considered a surface feeder. Weeding is therefore crucial for the establishment of plants. Weeding must be done often during the first year of planting. Subsequently, two to three rounds of hand weeding at the plant base in May, September, December/January and slash weeding in the interspace are advised, depending on the severity of weed development (Ankegowda *et al.*, 2022). Weeding can be done with mechanical weed cutters. These integrated, science-based practices have substantially improved the sustainability and productivity of cardamom ecosystems.

4.4.1 Irrigation and water management

Cardamom is highly sensitive to soil moisture deficits, particularly during panicle initiation (February–April) and capsule development (June–October). In the Western Ghats, increasing climate variability has intensified moisture stress, necessitating precision irrigation strategies. Recent studies at IISR, RS, Appangala (2021–2023) revealed that drip irrigation at 80% crop evapotranspiration (ETc) enhanced yields by 20–30% and conserved 40% water compared to conventional basin irrigation. Microsprinkler systems, maintaining higher humidity, promoted better spike emergence and reduced thrips incidence by about 25% (Srinivasan *et al.*, 2024). Fertigation through drip systems ensured precise nutrient delivery, improving nutrient-use efficiency and reducing losses. Irrigation scheduling based on ETc, determined using the Penman–Monteith equation, is now widely adopted. Additionally, soil-moisture-based scheduling (triggered when available moisture falls below 60%) and plant-based indicators such as leaf water potential (−0.8 MPa) are gaining traction. Although not widely studied in cardamom, use of smart sensors and automated irrigation controls as adopted in other annual crops could be a major advancement toward real-time precision irrigation management in cardamom ecosystems.

4.4.2 Nutrient management

Cardamom, which is a perennial with a thin root base and a long flowering period, requires balanced nutrition to ensure optimum yield. Excess nitrogen application helps vegetative development, but N, P, K and Mg deficiency cause leaf chlorosis, necrosis, poor spike formation and affect yield.

Recommended fertilization for irrigated small cardamom is $N:P_2O_5:K_2O$ at 75:75:150 kg ha^{-1} yr^{-1}, applied in three splits (May, September and January); rainfed systems receive 60% of this dose. Neem-coated or slow-release urea minimises leaching, while integration with organic sources such as FYM or vermicompost (2–5 t ha^{-1}) sustains soil organic matter. Phosphorus (75 kg P_2O_5 ha^{-1}) is vital for root development, and its availability improves with PSB or rock phosphate (200–250 kg ha^{-1}). Potassium (150 kg K_2O ha^{-1}), supplied through Muriate of Potash (MOP) or wood ash, regulates water balance and capsule quality; foliar sprays (1–2% K) enhance panicle growth and capsule filling.

Micronutrient deficiencies of Zn, B and Mg are prevalent in the Western Ghats' red soils. Foliar applications of 0.25% $ZnSO_4$ and 0.1% H_3BO_3 twice a year improve spike length and capsule set, while dolomite (500 kg ha^{-1} every 2–3 years) corrects Mg deficiency and maintains the soil pH. Foliar spray of 0.5% $FeSO_4$ prevents chlorosis in acidic soils. Integrated nutrient management involving Azospirillum, PSB and Arbuscular Mycorrhizal Fungi (AMF), particularly *Glomus fasciculatum*, enhances nutrient uptake and drought tolerance. Green manuring with legumes contributes 40–50 kg N ha^{-1} annually, reinforcing soil fertility and resilience (Thankamani *et al.*, 2024).

4.4.3 Fertigation and precision nutrient management

Drip-irrigation coupled with fertigation system offers accurate fertilizer delivery resulting in improved nutrient-use efficiency, and reduces leaching losses in high-rainfall zones (Srinivasan *et al.*, 2024). Adoption of drip-fertigation systems reduces 25% of the fertilizer usage that would typically contain 75% of the recommended fertilizer rates. The primary advantage assisted with the drip-fertigation, wherein nutrients would be delivered in 8–10 split doses during the active growing season. Soluble fertilizers recommended for cardamom cultivation are urea, potassium nitrate and monoammonium phosphate. By combining drip fertigation with biofertilizers can enhance crop yields by 20–30%, by reducing chemical fertilizer use by 25–30%. Thus, fertigation technique improves crop quality attributes such as capsule size and essential oil content. Recent developments in precision agriculture are also being used to hill spices. Sensors that measure soil moisture, electrical conductivity and canopy temperature enable data-driven irrigation and fertilization scheduling.

Cardamom plantations can efficiently use drones for monitoring, surveillance and decision support. However, there is very little utilization of drones for spraying operations on cardamom crops. The multi-tiered canopy that cardamom plants develop when cultivated in deep shadow limits the effective penetration of spray droplets to the lower plant portions, where pests and illnesses are frequently found. Additionally, the majority of cardamom plantations are situated in forested, steep and undulating terrain, which makes drone navigation challenging and raises the possibility of signal loss and tree accidents. Drone flight windows are further restricted by high humidity, frequent rains and misty conditions. Routine drone use is also less feasible due to regulations, high initial costs, the requirement for trained operators and the limited payload capacity, especially for small and dispersed holdings.

4.4.4 Shade management

Cardamom requires filtered light to achieve optimal photosynthesis and humidity levels. Excessive shade limits flowering, but open exposure causes more heat and wind damage. Therefore, more robust shade regulation practices are required in cardamom to obtain good yields. According to studies, tiny cardamom benefits from 40–60% light interception (Sivaraman *et al.*, 2023). Under 80% shade, vegetative growth is rich but flowering is poor, while under <30%, leaves burn and capsule set decreases. Whereas, cardamom plants with 50–75% shade with 300–900 mmol $m^{-2} s^{-1}$ photosynthetic photon flux density (PPFD), exhibited higher photosynthetic activity (Alagupalamuthirsolai *et al.*, 2019). However, the light intensity beyond these PPFD levels greatly affected cardamom growth and development. Therefore, selective lopping of shade trees such as *Artocarpus heterophyllus*, *Albizia falcataria* and *Grevillea robusta* prior to the monsoon season enables proper light penetration and aeration. Shade trees should have deep roots, a deciduous habit and non-allelopathic litter. Ankegowda and Krishnamurthy (1998) advocate indigenous trees such as *Ailanthus malabarica*, *Terminalia bellerica*, *Michelia champaca* and *Vernonia arborea*. Leguminous trees (*Albizia*, *Gliricidia* and *Erythrina*) enrich the soil with organic matter and nitrogen through leaf fall.

Cardamom plantations support a varied multi-story flora. Weeds that grow in abundance compete for nutrients and serve as breeding grounds for pests. The most typical approach is to manually weed three times a year (before, during and after the monsoon).

4.4.5 Climate-resilient crop management

With increasing climatic variability, adaptive management has become crucial for sustaining cardamom productivity (Adireddy *et al.*, 2025). Current research focuses on strategies that enhance resilience through mulch-based microclimate management, drip-cum-mist irrigation and drought-tolerant varieties such as IISR Manushree and IISR Kaveri. Agroforestry diversification with compatible tree species further stabilises microclimatic conditions. Simulations using decision support system for agro-technology transfer (DSSAT) models project a 10–15% yield decline by 2050 under existing practices, though precision irrigation and nutrient management could compensate for most losses (Thorp *et al.*, 2008). For better crop management, resource optimisation and climate-resilient agricultural planning, the DSSAT model facilitates crop growth simulation, yield prediction and climate effect assessment (Sannagoudar *et al.*, 2023).

Nutrient-diagnosis advancements now enable fine-tuned nutrient application. Established leaf standards – N (2.2%), P (0.25%) and K (2.3%) at full bloom aid in optimising fertilization. These leaf standards denote the optimal or critical nutrient concentration levels in a fully formed (index) leaf at a particular growth stage (full bloom) that serve as a benchmark for determining the nutrient status of plants. Portable chlorophyll meters (soil plant analysis development – SPAD) and leaf colour charts facilitate real-time nitrogen adjustments, reducing fertilizer use by 15% without yield reduction (Subbiah *et al.*, 2008).

Integrated crop management (ICM) modules synthesising these technologies have been standardised for smallholder adoption. Incorporating high-yielding clones, regulated drip fertigation, mulching, biofertilisers and integrated pest and disease management (IPDM) practices, field demonstrations in Kodagu and Idukki (2022–2024) indicated yield gains of 35–40% with a 25% decline in input costs (Ankegowda and Krishnamurthy, 2008). These integrated frameworks demonstrate promising pathways for climate-adaptive, resource-efficient cardamom production.

4.4.6 Intercropping systems

Intercropping increases land-use efficiency, diversifies farm income and contributes to improved pest control. The compatible intercrops (Table 4.1) include black pepper (*Piper nigrum*), which provides both shade and extra income; coffee (*Coffea arabica* or *C. robusta*), which is shade tolerant and commercially valuable; and legumes like *Pueraria* and *Dolichos*, which enrich the soil by increasing nitrogen levels and organic matter content.

Table 4.1. Recommended intercropping combinations for cardamom.

Intercrop	Planting pattern	Benefits
Black pepper	Along rows of cardamom	Shade + additional income
Coffee	Along boundary rows	Diversified income
Legumes	Between cardamom rows in the initial years	Soil fertility improvement

4.5 Pest and Disease Management in Cardamom

Since cardamom cultivation primarily confined to humid and high-rainfall tropical climates, is more prone to biotic stresses combined with insect pests, diseases and nematodes. Untreated cardamom crops with excessive pest and disease infestation can lead to crop losses of up to 70% (Venugopal *et al.*, 2006; Nair *et al.*, 2019). The dense canopy of trees created by the perennial nature and humid microclimate encourages the occurrence of pests and diseases in cardamom plantations. Therefore, adoption of IPDM with a focus on ecological and sustainable pest and disease control techniques that prioritize biological control, resistant varieties and precise diagnostics would lead to efficient pest and disease management.

4.5.1 Major insect pests of cardamom

The major pests infesting cardamom include cardamom thrips (*Sciothrips cardamom* Ramk.), shoot borer (*Conogethes punctiferalis* Guenée), root grub (*Basilepta fulvicorne* Jacoby) and capsule borer (*Dichocrocis punctiferalis*) (Table 4.2). Thrips are considered the most destructive pest, particularly during dry periods when population outbreaks occur.

IPDM techniques with maintaining optimal shade (50–60%) along with organic manures application significantly enhances natural predators such as

Table 4.2. Major insect pests of cardamom and their management approaches (KAU, 2017).

Pest	Nature of damage	Management strategies
Thrips *Sciothrips cardamom*	Scraping of tender leaves, flowers and capsules leading to silvering and shriveling	Application of neem oil (3%), *Beauveria bassiana*, or spinosad; maintenance of shade and humidity. Quinolphos 25 EC @ 0.05% or Dimethoate 30 EC @ 0.05% spray during rainless period
Shoot borer *Conogethes punctiferalis*	Larvae bore into pseudostems causing drying of central shoots	Pheromone traps, biological control (*Trichogramma chilonis*), pruning of infested shoots Quinolphos 25 EC @ 0.05% spray
Root grub *Basilepta fulvicorne*	Feeding on roots and rhizomes causes yellowing and wilting	Soil application of neem cake (250 kg/ha), *Metarhizium anisopliae*, crop sanitation
Capsule borer *Dichocrocis punctiferalis*	Larvae feed on immature capsules causing premature drying	Regular field inspection, neem extract sprays, light traps, 0.05 % Quinolphos 25 EC spray

Chrysoperla carnea, *Coccinella septempunctata* and *Oecophylla smaragdina* (Jacob *et al.*, 2020). Shanshank *et al.* (2015) demonstrated that use of pheromone-based mass trapping of shoot borers reduced pest incidence by 40–50% in large plantations. Application of Quinolphos 25 EC @ 0.05% spray reduces the pest incidence in cardamom significantly.

4.5.2 Nematode infestations

The major soil-borne nematodes that affect cardamom yields significantly are root-knot nematode (*Meloidogyne incognita*) and lesion nematode (*Pratylenchus coffeae*). These nematodes impair plant vigour significantly resulting in stunted growth and lower yields in cardamom. Poor or improper drainage under continuous monocropping of cardamom further increases the nematode activity. Integrated nematode management (INM) techniques were found to be alter nematode population in soil by enhancing crop yields. Use of biocontrol agents such as *Trichoderma harzianum* and *Paecilomyces lilacinus* at 25 g per clump enhances crop yield through minimizing nematode population. Growing of biofumigant crops like mustard and marigold as inter-row crops, soil solarizing nursery soil for roughly 30 days and adding 1 kg of neem cake per pit during planting significantly reduces nematode population. In a study at IISR (2021), it was found that using *P. lilacinus* and neem cake collectively reduced 78% nematode populations compared to untreated plots.

4.5.3 Major diseases of cardamom

Katte disease, caused by the cardamom mosaic virus (CdMV), is one of the most severe diseases, resulting in yield losses of up to 90%. The virus is spread by the banana aphid (*Pentalonia caladii*). Typical signs include stunted growth, chlorotic streaks and a mosaic pattern on leaves. Management measures include using virus-free planting material from indexed nurseries. The primary management techniques are rouging and eradication of diseased plants, and vector control with pesticides or neem-based formulations. The adoption of resistant varieties like Appangala 2 and IISR Vijetha effectively provide good returns to the farmers. Collectively, use of advanced biotechnological techniques like RNA interference (RNAi) and CRISPR/Cas9-based resistance to be extensively studied to develop virus-resistant cardamom lines (Khan *et al.*, 2022).

Capsule rot disease caused by *Phytophthora meadii* need to be controlled under high rainfall regions receiving >2500 mm. The pathogen targets fragile leaves, panicles and capsules, causing rotting and falling (Fig. 4.1a). Management techniques such as better drainage system and shade regulation were important at all the crop stages to control pest and diseases. Application of *Trichoderma viride* (10^6 cfu g^{-1}) along with farmyard manure is significantly affecting disease pressure in cardamom. In the regions with higher disease pressure, preventive spray of 1% Bordeaux mixture or 0.3% copper oxychloride is always advisable. Use of *Pseudomonas fluorescens* as a bioprotectant

Fig. 4.1. Diseases of cardamom (a) Capsule rot; (b) Rhizome rot.

enhances yield through lower disease pressure (Sabu and Menon, 2020). The microbial inoculation–chemical spray combination of metalaxyl (0.2%) or *T. harzianum*-enriched compost along with removal of disease affected tillers and disease-free organic debris as mulch is widely adopted by the farmers. Emerging pathogens such as *Colletotrichum gloeosporioides* (leaf blight) and *Lasiodiplodia theobromae* (panicle dieback) have been reported recently under changing climatic conditions (Nnadi and Carter, 2021). Climate change-driven fluctuations in humidity and temperature alter the pest and pathogen complex, demanding continuous monitoring and adaptive management.

Another major disease of cardamom is rhizome rot, caused collectively by *Pythium vexans* and *Rhizactonia solani*. The symptom appears as yellowing in leaves and the collar region becomes brittle, which leads to lodging of the plant (Fig. 4.1b). This disease also affects the young plants and leads to huge losses in cardamom production. Use of biocontrol agents in nursery mentioned in the nursery management section helps protect the plants from the pathogens. Apart from that, application of the fungicide mancozeb is found to be effective in controlling the severity and spread of the pathogens (Thomas *et al.*, 1997).

4.6 Handling, Processing and Quality Enhancement in Cardamom

Postharvest management is important as that of production technologies in the value chain of cardamom that determines capsule quality, market price and essential oil content. Cardamom is a highly perishable spice with delicate aroma and volatile oil content that deteriorates rapidly under improper handling, storage, or processing conditions. In the recent past, scientists have made great achievements in developing research and technological advancements that improved harvesting techniques, storage protocols and quality assessment, thereby enhancing both domestic and export competitiveness.

4.6.1 Harvesting of cardamom

Owing to the uneven ripening of the capsules, cardamom harvesting requires multiple pickings. Therefore, harvesting should be limited to physiologically matured capsules in order to achieve the desired litre weight, colour and market quality. Harvesting of fully ripened capsules are collected manually or hard picking, where semi-mature capsules are also harvested, depending on labour availability. The harvester should proceed from the base to the apex of each panicle by easily detaching capsules without pulling (Nimisha *et al.*, 2021). Fruit maturation requires approximately 110 days after flowering, and capsules are optimally harvested at physiological maturity, when the seeds have turned brown to black. Distinct physical and biochemical changes occur during physiological maturity that signifies the stage for perfect harvesting of cardamom capsules. Maintaining low seed moisture content with better dry matter content to be harvested when capsules turn from a dark green to a lighter green or yellowish hue. During this phase, the biosynthesis of essential oils stabilizes and the seeds become firmer and darker, resulting in improved flavour and aroma. These modifications ensure improved postharvest quality, storability and market value by reflecting the conclusion of seed development. Harvesting at this stage maximizes essential oil content and overall produce quality while minimizing losses from pest damage and capsule breaking. Over ripening must be avoided, as it increases losses due to rodent damage, natural shedding and capsule splitting during curing, all of which adversely affect both yield and quality. Improper harvesting practices lead to the deterioration of green colour and a consequent reduction in market value.

4.6.2 Curing

A vital postharvest operation after harvesting is curing that determines quality. The harvested capsules contain 70–80% moisture; hence, timely and scientific curing is essential. Curing starts when the capsules turn light-green stint with bold size at maturity. Immature capsules become shrivelled during drying, while over-matured capsules develop pale or yellowish colour after drying. After harvest, the capsules are cleaned thoroughly with water to avoid contamination by external matter. Sun drying is the most accessible method of curing, where capsules spread in thin layers on drying yards for 3–5 days. Direct sunlight exposure during sun drying cause bleaching and rapid chlorophyll degradation. Furthermore, uneven drying can cause fungus during storage, which makes sun-cured cardamom less desirable in high-end markets. To overcome these, artificial curing with flue pipe dryers and electrical dryers have gained importance. These dryers ensure controlled temperature and airflow, resulting in uniform drying and improved colour retention. Capsules are dried at 40–55°C for 18–36 h to retain bright green colour along with essential quality attributes with these dryers. The controlled heating inactivates enzymes responsible for chlorophyll degradation, while moisture removal helps maintain structural integrity. Concurrently, blanching in hot water, use of sodium carbonate for capsule dipping and modified-atmosphere drying are being followed in few cardamom production regions.

4.6.3 Storage and packaging

Proper storage minimizes loss of essential oils, colour degradation and microbial spoilage. Cardamom with more than 20% moisture (hygroscopic), should be stored under low temperature (15–20°C) and low humidity (<60%) in airtight containers (Table 4.3). Vacuum-sealed or nitrogen-flushed packaging is used for long-term storage and export, while perforated polypropylene bags or jute sacks are used for short-term storage with regular ventilation. Mehyar *et al.* (2014) found that under poor storage conditions over a period of more than six months, a 10–15% decrease in essential oil content was observed, while under controlled storage conditions, 95–98% of the essential oil content was retained.

4.7 Future Prospects, Climate Resilience and Technological Innovations

Cardamom is mostly grown in heavy rainfall, mild temperatures and humid microclimates of India, Guatemala and Sri Lanka. However, cardamom production is greatly hampered under these regions due to changing climate effects. Drought stress or waterlogging are well evident due to irregular rainfall patterns affecting flowering and capsule development. Temperatures above 35°C have a harmful effect on spike initiation and essential oil biosynthesis. Warmer winters can contribute to an increase in pest and disease problems, with thrips, shoot borers and fungal diseases in particular, able to survive well at these temperatures. Furthermore, soil deterioration caused by increased erosion and fertilizer loss reduces agricultural output.

4.8 Climate-resilient Cultivation Strategies

Molecular and marker-assisted breeding could help in developing resilient crop varieties characterized by high yields, stress tolerance and disease

Table 4.3. Recommended storage practices for cardamom.

Parameter	Recommendation
Moisture content	10–12%
Temperature	20–25°C
Relative humidity	60–65%
Packaging	Airtight containers, Modified Atmosphere Packaging if possible
Shelf life	12–18 months

resistance. Clonal and micropropagation techniques also help in developing virus-free, homogeneous and scalable planting materials. Drought mitigating techniques such as microirrigation and fertigation technologies enhance water-use efficiency with reduced water usage. Adoption of organic mulching significantly helps in conserving moisture and stabilizing soil temperatures. Adaptive shadow management, particularly with leguminous trees, manages the microclimate and humidity, with an optimal canopy cover of 50–60% minimizing heat stress and pest pressure while increasing photosynthesis. To maintain soil fertility in changing climates, application of organic amendments, green manures and biofertilizers are essential. Growing of legumes as intercropping and cover crops improve nitrogen fixation and help to minimize soil erosion. Development of integrated pest and disease management systems association with resistant cultivars, biocontrol agents and pheromone traps to provide more effective pest and disease control. Climate-based early warning systems enable prompt responses against pests and diseases. Sensor-based climate smart technology using drones, and AI systems improved real-time crop monitoring and decision making, which advances overall crop management. Finally, postharvest advancements such as sun dryers, modified environment packing and microencapsulation increase crop shelf life while preserving essential oil quality.

4.9 Conclusion

Over the past two decades, development of advance breeding lines has led to enhanced cardamom productivity with greater export potential. Sustainable cardamom production and improved quality have been augmented through germplasm characterization, disease-free planting stocks and high-yielding varieties. However, long crop cycles and low genetic diversity hindered cardamom cultivation. Future programmes must connect advanced technologies such as micropropagation, microirrigation, IPDM and digital decision-support systems to develop climate-resilient technologies adaptable to diverse agroecologies. Propagation begins in the nursery and the focus should be on disease-free, genetically pure and physiologically vigorous plants, supported by biotechnology for fast multiplication of advanced cardamom lines. As climate change is threatening cultivation with pests, diseases and soil problems, innovative cultivation methods are therefore crucial for crop resilience. A sustainable future depends on integrating genetic improvement, precision farming, biotechnological advances and value addition, alongside investments in research, infrastructure, farmer training and digital tools to ensure global competitiveness.

5

Bioactive Constituents of Cardamom: Therapeutic Potential and Medicinal Properties

Kaliyaperumal Ashokkumar[1]*, M. Murugan[2], M. Alagupalamuthirsolai[3] and Saradhadevi Muthukrishnan[4]

[1]School of Agriculture and Animal Sciences, Gandhigram Rural Institute-Deemed to be University, Gandhigram, Dindigul, Tamil Nadu, India
[2]Cardamom Research Station, Kerala Agricultural University, Pampadumpara, Kerala, India
[3]ICAR-Sugarcane Breeding Institute, Coimbatore, Tamil Nadu, India
[4]Department of Biochemistry, Bharathiar University, Coimbatore, Tamil Nadu, India

5.1 Cardamom Essential Oil Yield and Extraction Methods

Cardamom essential oil (CEO) is a high-value product obtained primarily from matured fruits of cardamom. It is isolated using various techniques that have different effects on its yield and chemical composition. Traditional methods such as steam distillation and hydro-distillation remain widely used due to their ease and safety, though they typically yield lower essential oil percentages. In contrast, solvent extraction offers yields up to 7.3%, though it may introduce solvent residues requiring further purification (Sontakke *et al.*, 2018). In a recent study involving 22 diverse cardamom accessions from Southern India, essential oil yields ranged from 4.5% to 9.5% highlighting the genetic variability and potential for genotype selection in breeding programmes (Ashokkumar *et al.*, 2021).

For laboratory testing, hydro-distillation is the most convenient method for obtaining CEO, while for commercial production, steam distillation is the

*Corresponding author: k.ashokkumar@ruraluniv.ac.in

preferred method to account for larger samples and a superior volatile profile of CEO. Recent innovations include ultrasound-assisted extraction, microwave-assisted extraction and pressurised liquid extraction, which enhance efficiency by disrupting cell walls and improving essential oil release. These advanced methods increase yield and preserve thermolabile compounds, making them suitable for high-quality essential oil production (Katekar *et al.*, 2023). The supercritical carbon dioxide ($scCO_2$) extraction is gaining momentum for extraction of active constituents used for pharma and nutraceutical applications. The procedure uses carbon dioxide above its critical point (critical temperature of 31.1°C and critical pressure of 7.38 MPa), where the gas acts as a liquid-like solvent but flows as a gas-like fluid. The main advantage of using $scCO_2$ is obtaining the extract, which is non-toxic and residue-free. It is also used for selective extraction of phytochemicals by varying the temperature and pressure of the system. Hamdan *et al.* (2008) studied the extraction of cardamom oil using $scCO_2$ and sub-critical propane vs the traditional hydro-distillation. The results show that sub-critical propane is superior to $scCO_2$ for industrial extraction of high-quality cardamom essential oil in terms of yield, speed and aroma profile, while still being a 'green' solvent process, i.e. propane is easily recovered and recycled. Even though $scCO_2$ remains attractive when an absolute absence of any hydrocarbon residue is mandatory (e.g. certain food/pharma regulations). As mentioned before, the choice of method depends on the intended application, desired purity and available infrastructure, with modern techniques increasingly favoured in pharmaceutical and food industries.

Moreover, the yield of essential oil varies significantly depending on factors such as variety, capsule maturity, soil type, origin and environmental conditions (Alagupalamuthirsolai *et al.*, 2024). The yield of essential oil from various parts and various extraction methods of cardamom is summarized in Table 5.1.

5.2 Chemistry, Preparation Methods and Stability of Encapsulated CEO

The volatile compounds of CEO, particularly α-terpinyl acetate, 1,8-cineole and limonene, tend to evaporate and may undergo oxidative degradation, which significantly limits their shelf life and effectiveness. Therefore, selected encapsulation matrices such as cyclodextrins, maltodextrin, gum arabic, whey protein isolates and bio-polymers are used, which interact with the CEO molecules via hydrogen bonds and protect them from external stress factors such as light, oxygen and heat (Ming *et al.*, 2025).

Several preparation methods have been optimised to achieve efficient encapsulation. Nanoemulsification, achieved through ultrasonication or high-pressure homogenisation, produces CEO droplets of less than 200 nm, ensuring uniform dispersion and enhanced bioavailability (Souza *et al.*, 2022).

Table 5.1. Essential oil yields obtained from different parts of the cardamom plant.

Process method	Plant part used	Oil yield (%)	References
Solvent extraction	Whole capsules	7.3	Sontakke *et al.* (2018)
Steam distillation	Whole capsules	2.6	Sontakke *et al.* (2018)
Hydro-distillation	Capsules (pre-ripe)	1–5	Morsy (2015)
Hydro-distillation	Whole capsules	2.2–9.6	Ashokkumar *et al.* (2020a, 2020b, 2020c, 2021)
Hydro-distillation	Whole capsules	2.4–3.1	El Sayed *et al.* (2022)
Hydro-distillation	Seeds	6.9	Andriza *et al.* (2022)
Hydro-distillation	Leaves	0.26–0.33	Pujiarti and Kusumadewi (2020)
Hydro-distillation	Rhizomes	0.06	Pujiarti and Kusumadewi (2020)
Vacuum distillation	Crushed capsules	3.5	Kiralan *et al.* (2023)
Cold pressing	Seeds	1.2	Kiralan *et al.* (2023)
scCO_2 extraction	Seeds	6.0	Daood (2008)

Spray-drying techniques, using carriers such as maltodextrin, gum arabic or modified starch, yield dry powders with improved handling and storage stability. Biodegradable polymeric nanoparticles, including chitosan and poly lactic-co-glycolic acid (PLGA), have been widely used to entrap CEO, offering controlled release and targeted delivery (Jamil *et al.*, 2016). Coacervation methods have also been employed to encapsulate essential oils, producing microcapsules with high encapsulation efficiency and improved oxidative stability (Souza *et al.*, 2022). More recently, electrospinning has been explored as a novel encapsulation technique, where CEO is incorporated into nanofibers, providing sustained release and enhanced antimicrobial activity (de Oliveira *et al.*, 2025).

Encapsulation markedly improves the stability of cardamom essential oil. Studies have reported that encapsulated CEO retain more than 80% of their antioxidant activity after prolonged storage, compared to less than 40% in free essential oils (Abdullah *et al.*, 2021; Savan, 2023). Encapsulation also reduces volatility, delays oxidation and enhances solubility in aqueous systems, thereby broadening its usage in beverages, functional foods and pharmaceutical formulations (Al-Ismail *et al.*, 2015). For example, the CEO encapsulated in β-cyclodextrin complexes demonstrated significantly higher retention of monoterpenes during thermal processing compared to free essential oil. Similarly, spray-dried CEO powders maintained antimicrobial activity for up to 6 months of storage, whereas free essential oil lost more than half of its activity within 2 months (Kumari *et al.*, 2014). These advances underscore the importance of encapsulation as a critical step in harnessing the full potential of CEO for industrial and therapeutic use.

5.3 Bioactive Components of CEO

The chemical profiles of the CEO exhibit significant regional and varietal variation. In the Guatemalan seed oil, 1,8-cineole (55.4%) and α-terpinyl acetate (18.2%) dominate, followed by limonene (6.3%) and terpinene (4.1%), suggesting a cineole-rich chemotype (Kiralan *et al.* 2023). In a study by Ashokkumar *et al.* (2021), 22 accessions from Southern India showed a significant variability in the essential oil composition, with α-terpinyl acetate ranging from 29.9% to 61.3% and 1,8-cineole from 15.2% to 49.4%. Furthermore, major essential oil compositions from various origins and plant parts of cardamom are summarized and presented in Table 5.2.

The essential oil yield and its composition is also affected by pest infestation, especially by cardamom thrips, caused by *Sciothripscardamomi*, which leaves visible scars over the capsules, reducing the visual appearance of the capsule. A study by Muthusamy *et al.* (2025) categorised the infestation into six scales (Scale I–VI) based on the extent of damage to the capsules. They reported that the content of essential oil started declining after scale III and observed 53.5% CEO yield loss at scale VI. Moreover, the content of 1,8-cineole, which imparts camphorous flavour, starts increasing and the content of α-terpinyl acetate, which imparts sweet flavour, starts decreasing with increasing thrips damage.

Cardamom capsules are rich in bioactive flavonoids, including catechin (281.8 $\mu g\ g^{-1}$), myricetin (18.6 $\mu g\ g^{-1}$), kaempferol (8.7 $\mu g\ g^{-1}$) and quercetin (4.9 $\mu g\ g^{-1}$), which contribute to its antioxidant potential and therapeutic

Table 5.2. Essential oil compositions in different places of origin and plant parts of cardamom.

Origin	Plant parts	Major chemical constituents (%)	References
Southern India	Fruits (capsules)	α-terpinyl acetate (29.9–61.3%), 1,8-cineole (15.2–49.4%), linalool (3.2–6.1%), sabinene (2.1–4.8%)	Murugan *et al.* (2019); Ashokkumar *et al.* (2021)
Guatemala	Seeds	1,8-cineole (55.4%), α-terpinyl acetate (18.2%), limonene (6.3%), terpinene (4.1%), sabinene (3.7%)	Kiralan *et al.* (2023)
India (Malabar type)	Seeds	α-terpinyl acetate (20–55%), 1,8-cineole (20–60%), sabinene (2–6.2%), linalool (-3–5.8%)	Jadav and Mehta (2018)
India	Fruits (capsules)	α-terpinyl acetate (61.65–68.19%), cineole (7.23–11.76%), limonene (4.5%), terpinene (3.2%)	Sontakke *et al.* (2018)

relevance. Among carotenoids, lutein is the most abundant at 2.4 μg g^{-1}, followed by β-carotene at 0.5 μg g^{-1} (Ashokkumar *et al.*, 2020c). These phytochemicals are vital in mitigating oxidative stress and supporting cellular health. Recent investigations have highlighted that phenolic constituents of cardamom, such as flavonoids, alkaloids, terpenoids and anthocyanins, exhibit pharmacological activities that may aid in the treatment of cardiovascular, pulmonary, renal and respiratory disorders (Vaidya and Rathod, 2014). Flavonoids and phenolic acids further enhance its pharmacological profile, providing antioxidant protection that may help mitigate oxidative stress-related conditions, such as diabetes, cardiovascular disease and neurodegeneration.

5.4 Therapeutic Potential and Medicinal Properties

Cardamom has been valued for millennia across cultures both as a culinary spice and as a powerful medicinal herb. Across different continents, cardamom remains a household remedy for respiratory congestion, post-partum recovery and as a warming carminative, most famously infused into Arabic/Turkish coffee to neutralise caffeine's harshness and enhance digestion. Several *in vitro* and *in vivo* studies done on CEO and cardamom extracts confirmed their broad spectrum of biological activities, including antioxidant, antimicrobial, anti-inflammatory, anticancer, antidiabetic, antiviral, gastroprotective, cardioprotective, neuroprotective and several other pharmacologically relevant effects. Also, a diagrammatic representation of the therapeutic and health benefits of cardamom is shown in Fig. 5.1.

5.4.1 Antioxidant activity

Cardamom exhibits potent antioxidant activity through its extracts, essential oil and bioactive compounds, including 1,8-cineole, α-terpineol, linalool and α-terpinyl acetate. These compounds act via free radical scavenging, metal chelation and enhancement of endogenous antioxidant enzymes, including superoxide dismutase (SOD), catalase (CAT) and glutathione peroxidase (GPx). Essential oil obtained via hydro-distillation exhibited strong antioxidant capacity *in vitro*, with DPPH (2,2-diphenyl-1-picrylhydrazyl) inhibition reaching 79.48% at optimised extraction conditions using instant controlled pressure drop (ICPD) technology coupled with sonication (Raissa *et al.*, 2020). Additionally, green cardamom essential oil showed IC_{50} values of 45.3 μg/ml and 61.2 μg/ml in the DPPH and ABTS (2,2'-azino-bis(3-ethylbenzothiazoline-6-sulfonic acid) assays, respectively, and its major compounds demonstrated favourable binding affinities for antioxidant-related enzymes in molecular docking studies (Alam *et al.*, 2021). Elguindy *et al.* (2016) demonstrated that cardamom extract at 200 mg/kg significantly decreased hepatic MDA levels while increasing reduced glutathione (GSH) concentrations. Also, key

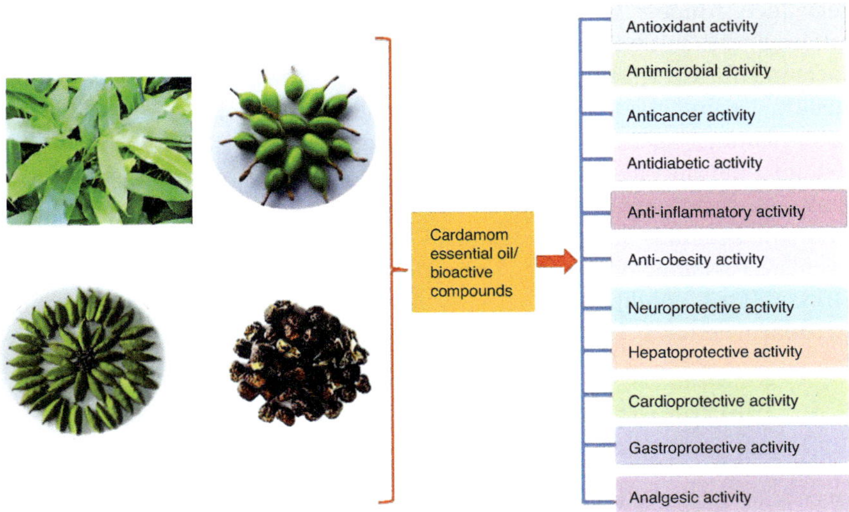

Fig. 5.1. Diagrammatic representation of the therapeutic potential and health benefits of cardamom.

antioxidant enzymes, including CAT, SOD, GPx, glutathione reductase and glutathione-S-transferase (GST), were all elevated in treated groups compared to controls. These effects contribute to its therapeutic potential against oxidative stress-related conditions, reinforcing its pharmacological relevance in natural product-based interventions.

5.4.2 Antimicrobial activity

CEO exhibits broad-spectrum antimicrobial activity, primarily attributed to its rich composition of bioactive compounds such as 1,8-cineole, α-terpinyl acetate and linalool. Noktehsanj *et al.* (2020) reported that the CEO demonstrated inhibitory effects against *S. aureus*, *E. coli* and *S. cerevisiae*, with minimum inhibitory concentrations (MICs) ranging from 12.5 to 25.0 mg/ml due to disruption of microbial membranes and metabolic pathways. In another study, CEO effectively inhibits the growth of *S. aureus*, *S. typhimurium*, *C. albicans* and *E. coli*, with MICs ranging from 5 to 10 mg/ml depending on the strain. Furthermore, it is demonstrated that the CEO effectively inhibited quorum sensing in *Chromobacterium violaceum* and inhibition of biofilm formation up to 85.59% in *E. coli* O157:H7 and 100% in *S. typhimurium*, suggesting its potential in anti-virulence strategies (Abdullah *et al.*, 2017, 2021). Jamil *et al.* (2016) demonstrated that CEO encapsulated into chitosan nanoparticles by ionic gelation method has resulted in more than 90% encapsulation efficiency and demonstrated strong antimicrobial activity against extended-spectrum β-lactamase–producing *E. coli* and methicillin-resistant

Staphylococcus aureus without having adverse effects on normal cell line. The study showed the selective toxicity against bacteria by CEO nanoparticle. These findings collectively support using the CEO as a natural antimicrobial agent in food preservation and therapeutic applications.

5.4.3 Anti-inflammatory activities

The CEO and its bioactive constituents have demonstrated promising anti-inflammatory properties across various experimental models. Alam *et al.* (2021) reported that 1,8-cineole and terpinen-4-ol exhibited strong molecular docking interactions with COX-2 and 5-LOX enzymes, showing binding affinities ranging from −6.5 to −7.2 kcal/mol, comparable to ibuprofen. Notably, cardamom oleoresins exhibited greater inhibition than the essential oil, attributed to their higher phenolic content, reinforcing the role of phytochemical composition in modulating anti-inflammatory efficacy. In a recent study, mice were orally administered cardamom extract at a dose of 200 mg/kg daily for 15–16 days, while acetaminophen toxicity was induced with a single intraperitoneal injection of 300 mg/kg on day 16. Pre-treatment with cardamom extract at 200 mg/kg significantly attenuated APAP (N-acetyl-p-aminophenol)-induced hepatic injury by lowering serum liver function markers, reducing oxidative stress and inflammation and enhancing antioxidant enzyme activities. These findings highlight that 200 mg/kg cardamom extract offers protective efficacy against APAP-induced acute liver failure through activation of the Nrf2/HO-1/NQO-1 pathway and suppression of inflammatory and apoptotic mediators (Alkhalifah, *et al.*, 2022). Moreover, clinical investigations in type 2 diabetic patients revealed that supplementation with 3 g/day of cardamom powder for 10 weeks led to significant reductions in pro-inflammatory cytokines, including TNF-α, IL-6 and C-reactive protein, compared to placebo controls (Aghasi *et al.*, 2018).

Cardamom supplementation has emerged as a promising natural intervention in metabolic disorders, particularly diabetes, through its ability to modulate key molecular targets associated with inflammation and oxidative stress. Among these, sirtuin 1 (Sirt1) plays a pivotal role in regulating glucose metabolism, insulin sensitivity and mitochondrial function. Activation of Sirt1 enhances pancreatic β-cell survival, improves insulin secretion and promotes glucose uptake in peripheral tissues, thereby alleviating hyperglycaemia and metabolic dysfunction. In a randomised, placebo-controlled, double-blind trial involving 80 pre-diabetic subjects, daily supplementation with 3g of cardamom powder for 8 weeks significantly increased Sirt1 expression compared with placebo, highlighting its potential in restoring metabolic homeostasis (Daneshi-Maskooni *et al.*, 2018).

Equally important are systemic inflammatory markers such as high-sensitivity C-reactive protein (hs-CRP) and the hs-CRP:IL-6 ratio, which serve as sensitive indicators of chronic low-grade inflammation in diabetes. Elevated hs-CRP levels are strongly correlated with insulin resistance, endothelial dysfunction and increased risk of cardiovascular complications in diabetic patients. Cardamom supplementation resulted in a significant reduction in serum hs-CRP and the hs-CRP:IL-6 ratio ($p = 0.008$), suggesting attenuation of systemic inflammation and improved inflammatory balance.

5.4.4 Cardioprotective activity

Cardamom extracts (CE) and CEO exhibit promising cardioprotective effects. In a preclinical study, Shahidullah *et al.* (2017) demonstrated that a maceration extract of cardamom, administered at doses of 100–500 mg/kg body weight, significantly mitigated doxorubicin-induced cardiotoxicity in rats by lowering serum levels of SGOT, SGPT, LDH and total cholesterol, while improving HDL levels, suggesting membrane stabilisation and lipid profile modulation in cardiac protection. Additionally, the antioxidant and anti-inflammatory properties of cardamom essential oil mediated through inhibition of lipid peroxidation and suppression of inflammatory cytokines further contribute to its cardioprotective potential (Alam *et al.*, 2021). The synergistic action of its terpenoid-rich profile supports its therapeutic relevance in managing oxidative stress and inflammation-linked cardiovascular disorders.

5.4.5 Neuroprotective activity

Limited studies have investigated CEO and its bioactive components for neuroprotective effects in preclinical models of Alzheimer's disease (AD). Auti *et al.* (2019) investigated its efficacy in aluminium chloride-induced neurotoxicity in rats and reported significant improvements in cognitive and behavioural parameters, alongside reduced oxidative stress and acetyl cholinesterase (AChE) activity. Histopathological analysis revealed diminished amyloid β plaque formation and enhanced brain-derived neurotrophic factor (BDNF) expression, indicating its potential in modulating AD pathology. Similarly, Ashour *et al.* (2024) examined the synergistic impact of nanoparticle-encapsulated cardamom essential oil (NCEO) and swimming exercise in aluminium-induced AD rat models. Their findings demonstrated improvements in biochemical and histological markers, reinforcing the role of NCEO in mitigating oxidative damage and supporting neuronal integrity. These studies suggest the therapeutic promise of cardamom essential oil and its nanoformulations in the management of neurodegenerative diseases.

5.4.6 Anticancer activity

In vitro studies have demonstrated that CEO significantly reduces cell viability in triple-negative breast cancer (MDA-MB-231) and kidney cancer (HEK 293) cell lines, with IC_{50}: 96.95 μl/ml and 157.44 μl/ml, respectively (Ahamad, 2024). Additionally, cardamom bioactive components such as cardamonin and indole-3-carbinol have been shown to modulate apoptosis-related pathways and overcome drug resistance in cancer cells (Qiblawi *et al.*, 2020; Bano *et al.*, 2024). Rahman *et al.* (2017) reported that cardamom powder supplementation (100–200 mg/kg/day) in obese rats reduced hepatic oxidative stress and inflammation, thereby lowering cancer risk factors. Furthermore, Jamil *et al.* (2016) highlighted that nano-encapsulated CEO with enhanced stability showed good cytotoxic efficacy against colon cancer cell lines and produced no toxic effects in normal colon cell lines, suggesting improved delivery and non-toxicity. Collectively, these findings from cell culture assays, molecular docking, animal models and encapsulation studies underscore the therapeutic potential of cardamom as a natural anticancer agent.

5.4.7 Antidiabetic activity

The CEO's bioactive constituents, including 1,8-cineole, α-terpinyl acetate, linalool and flavonoids, have been widely recognized for their dual role in modulating inflammation and glucose metabolism. These compounds enhance insulin sensitivity, regulate glucose uptake and reduce oxidative stress in pancreatic tissues, thereby contributing to improved glycemic control. The anti-inflammatory activities of cardamom are closely linked to its antidiabetic effects. Chronic low-grade inflammation is a hallmark of type 2 diabetes, contributing to insulin resistance and vascular complications. Clinical studies have demonstrated that daily supplementation with 3 g of cardamom powder for 10 weeks significantly improved inflammatory and endothelial markers in type 2 diabetic patients, suggesting a beneficial role in reducing vascular inflammation and improving endothelial function. However, no significant effect was observed on MMP-9 and CD163, indicating selective modulation of inflammatory pathways (Zahedi *et al.*, 2021). The macrophage scavenger receptor CD163 is considered a marker of anti-inflammatory macrophage activation. Its lack of modulation in this study highlights that cardamom's effects may be more pronounced in systemic inflammatory mediators rather than macrophage-specific pathways.

Similarly, another clinical trial reported that daily intake of 3 g of cardamom for 8 weeks in pre-diabetic subjects selectively reduced systemic inflammation and oxidative stress markers, particularlyhs-CRP and MDA (Kazemi *et al.*, 2017). hs-CRP is a sensitive biomarker of systemic inflammation, while MDA reflects lipid peroxidation and oxidative stress. Reduction of these markers is significant in alleviating diabetic symptoms, as systemic inflammation and oxidative stress are major contributors to insulin resistance, β-cell dysfunction and vascular complications. By lowering hs-CRP and MDA, cardamom supplementation not only attenuates inflammatory burden but also improves metabolic resilience, thereby offering protection against the progression of diabetes.

5.4.8 Hepatoprotective activity

Cardamom and its bioactive constituents exhibit robust hepatoprotective effects across diverse experimental models. Rahman *et al.* (2017) demonstrated that supplementation of cardamom powder at 1% in chow diet effectively restored liver enzyme levels and histological architecture in high carbohydrate high fat (HCHF) diet-induced obese rats. Similarly, Aboubakr and Abdelazem (2016) reported that aqueous cardamom extract at doses of 100 and 200 mg/kg (orally 7 days) significantly reduced serum AST, ALT, bilirubin and LDL, while elevating albumin and HDL levels, with marked histopathological protection against gentamicin-induced liver damage in rats. In a paracetamol-induced hepatorenal toxicity model in rats, Khattab *et al.* (2020) found that oral supplementation of CEO at 100 mg/kg for 14 days restored the activities of liver enzymes, total protein and enhanced total antioxidant capacity, preserving liver histology. These studies highlight cardamom's therapeutic potential for managing liver injury through biochemical restoration and antioxidant defence mechanisms.

5.4.9 Anti-obesity activity

Cardamom exhibits promising anti-obesity effects through its essential oil and ethanolic seed extracts, as demonstrated in both *in vivo* and *in vitro* models. In a recent study, a meta-analysis of seven clinical trials found that cardamom supplementation significantly reduced triglyceride levels compared with controls, and in three smaller studies, cardamom intake was associated with a significant increase in body mass index (BMI). However, no significant effects were observed on total cholesterol, LDL-cholesterol, HDL-cholesterol, fasting plasma glucose or body weight (Asbaghi *et al.*, 2020). These results collectively underscore cardamom's potential as a natural anti-obesity agent through modulation of lipid metabolism and adipocyte differentiation.

5.4.10 Miscellaneous activities

In a recent study, Mohemmad *et al.* (2024) demonstrated that a CEO loaded with nanostructured lipid carriers (CEO-NLC) significantly alleviated ethanol-induced gastric ulcers in rats at 300 and 600 mg/kg doses. CEO-NLC treatment reduced ulcer area, gastric acidity and MDA levels while enhancing total antioxidant capacity and anti-inflammatory cytokine IL-10. These effects were comparable to omeprazole (20 mg/kg), the standard reference drug. Additionally, CEO-NLC suppressed pro-inflammatory cytokines such as TNF-α and IL-6, indicating its role in modulating oxidative stress and inflammatory pathways in gastric mucosal protection. CEO exhibits bronchodilatory, mucolytic and anti-inflammatory properties that support respiratory health. Studies have shown that the CEO can inhibit microbial growth in the respiratory tract and modulate oxidative stress, thereby improving pulmonary function and reducing congestion (Noumi *et al.*, 2018).

5.4.11 Safety of cardamom essential oil

The Flavour and Extract Manufacturers Association (FEMA) is a US-based food industry trade group which employs the Threshold of Toxicological Concern (TTC) approach as a key component of its safety evaluation for natural flavour complexes (NFCs) like cardamom seed essential oil (FEMA 2241, CAS 8000-66-6). This method provides conservative exposure thresholds below which adverse health effects are unlikely, based on structural alerts for toxicity. It is applied to congeneric groups (chemically related constituents) and unidentified components after full characterization of the oil's composition. Based on per capita consumption in the US (annual volume of ~100–200 kg for flavour purposes), total essential oil intake is low at ~0.0004 mg/kg body weight/day, and the group-wise estimated intake is even a fraction of that value (Table 5.3).

All estimated dietary intakes fall substantially below TTCs, supporting GRAS affirmation for the intended flavour use of cardamom essential oil (Fukushima *et al.*, 2020).

If cardamom essential oil were used for a food supplement or nutraceutical, the amount of intake would be significantly higher than the quantity intended for flavour use. What would be the adverse effects of higher concentrations of cardamom essential oil consumption? A report by Hoberman (1989) showed that cardamom essential oil showed no mutagenic potential, testing negative in Ames assays using multiple strains of *Salmonella typhimurium* and in an Unscheduled DNA Synthesis (UDS) assay in rat hepatocytes at concentrations up to 50 mg/ml. However, in a reproductive/developmental toxicity study, female rats receiving cardamom essential oil by gavage at 375, 750, or 1500 mg/kg body weight/day for 7 days before mating and throughout mating, gestation, delivery and the first 4 days of lactation exhibited clinical

Table 5.3. TTC and estimated intake of cardamom essential oil compounds.

Congeneric group	Example constituents	TTC (μg/kg body weight/day)	Estimated intake (μg/kg bw/day)
Aliphatic/aromatic hydrocarbons	Sabinene, limonene	30	0.05–0.1
Cyclic ethers	1,8-cineole	30	0.1–0.2
Tertiary alcohols/ esters	α-terpineol, α-terpinyl acetate, linalool	30	0.15–0.25
Simple phenols (trace)	Eugenol derivatives	1.5	<0.01

signs of toxicity, including salivation, reduced motor activity and neurological effects such as twitches and tremors. Despite these maternal effects, no adverse impact was observed on mating performance, fertility, gestation length or parturition. Therefore, the no-observed-adverse-effect level (NOAEL) for reproductive parameters was established at the highest dose tested, 1500 mg/kg body weight/day. This amount is very high even for a nutraceutical application, which may use fraction of the quantity of the highest dose tested. The report conclusively showed that cardamom essential oil is safe for food, flavouring and nutraceutical usage. An EFSA (2019) report showed that cardamom essential oil used as an additive in animal feed also did not cause adverse effects. They concluded that the proposed level of 5 mg/kg in animal feed and 5 mg/l of water for drinking is considered safe for all animal species.

5.4.12 Conclusion

Cardamom extracts, essential oils and their diverse bioactive constituents represent a multifaceted phyto-therapeutic resource, distinguished by a rich composition of oxygenated monoterpenes such as 1,8-cineole, α-terpinyl acetate, linalool and α-terpineol. These compounds collectively underpin the spice's broad spectrum of pharmacological activities, ranging from antioxidant and anti-inflammatory effects to metabolic regulation. Advances in extraction methodologies and encapsulation technologies have further enhanced the bio-efficacy, stability and targeted delivery of cardamom-derived formulations, thereby expanding their relevance across pharmaceutical, nutraceutical and functional food applications. Taken together, these findings reinforce the therapeutic promise of cardamom essential oil and highlight its potential as a natural, sustainable and integrative approach to modern healthcare.

Quality Standards of Cardamom 6

R. Sivaranjani*

ICAR-Indian Institute of Spices Research, Kozhikode, Kerala, India

6.1 Grading in Cardamom

For marketing purposes, cardamom capsules are classified into different grades according to their size and colour. The premium grade of cardamom from India, i.e. Alleppey Green is slightly smaller in size than the highest grade quality of cardamom from Guatemala, i.e. Jumbo Green, nevertheless the flavour profile of Indian cardamom is far superior to its Guatemalan counterpart. In India, the quality standards of cardamom are fixed by the Spices Board of India. Alleppey Green Bold (AGB) with more than 7 mm pod size is traded primarily for export. The cardamom cultivated in the Kerala region is traded as Alleppey green, whereas the cardamom obtained from Karnataka is traded as Coorg green. The different grades of cardamom traded in the Indian market are explained in Table 6.1.

Compared to India, there are no extensive quality grades for Guatemalan cardamom. According to the 2014 report by the US Food and Drug Administration (USFDA) (Milian LS, 2014), the quality grades have been established for the export of Guatemalan cardamom (Table 6.2):

The other unspecified category of Guatemalan cardamom (www.tridge.com/market-guides/GT-cardamom) includes:

- 'Fancy Green Extra' has extra green pods.
- 'Fancy Green' with medium-sized green pods.
- 'Imperial Mixed Green' has large pale green pods.
- 'Mixed Green' category has pods of assorted colours.
- 'Mixed Green Split' includes medium-sized open green pods.
- 'Yellow Mixed' has medium/large closed yellow pods.

*Corresponding author: ranjanigop@gmail.com

Table 6.1. Grades of Indian cardamom.

Grade	Specifications	Size of capsules (mm)	Litre weight (g/L)
Alleppey Green Superior Extra Bold (AGSEB)	Uniform dark green; full, plump pods; no clipping; low defects (<2% shrivels/opens/ malformed)	8 or above	435
Alleppey Green Bold (AGEB)	Intense green; mostly full pods; minimal yellowing (<2% shrivels/opens/ malformed)	7–8	415
Alleppey Green Bold (AGB)	Deep green to green (<2% shrivels/opens/ malformed)	6–7	400
Alleppey Green Superior (AGS)	Green to light green; 90% full pods; low defects; no clipping (<2% shrivels/opens/ malformed)	5–6	385
Alleppey shipment green/ Alleppey green light	10–15% black and split capsules; 8 capsules and immature and shrivelled capsules	3.5–4.0	260–350
Coorg Green Extra Bold (CGEB)	Golden to light creamy in colour with no clipped or malformed capsules; no shrivelling in capsules	8 and above	450
Coorg Green Bold (CGB)	Creamy light greenish to greenish in colour; up to 3% immature or shrivelled capsules and up to 2% empty or malformed capsules are allowed	7.5–8	435
Coorg Green 1 – Superior	Up to 3% immature or shrivelled capsules and up to 3% empty or malformed capsules are allowed	6.5–7.5	415
Other grades of Coorg green (Motta green, shipment and light)	Up to 5% unclipped capsules; up to 3% immature or shrivelled capsules and up to 3% empty or malformed capsules are allowed	3.5–6.0	280–385
Bleached/white	Pale/white from sulfur bleaching; higher moisture tolerance; milder aroma	5–8.5 (mixed)	300–340
Open/splits	Yellow/mixed colour; >20% opens/ shrivels; ungraded mix	<5.5	<350
Mixed Green	Dried and mixed capsules of different varieties of *Elettaria cardamomum;* 2–3% immature or shrivelled capsules and up to 2–3% empty or malformed capsules are allowed; 10–12% blacks and splits are allowed	3.5–7	260–435

Ref: IS 1907:2023 (BIS, 2023)

Table 6.2. Export quality grades of Guatemalan Cardamom (Milian LS, 2024)

Grade	Specifications	Size of capsules (mm)	Litre weight (g/L)
Jumbo Green	Extra-large green small cardamom pods	8 and above	450
Imperial Best Green	Large green pods	6.7–7	415

6.2 Physical and Chemical Quality Standards in Cardamom

In addition to the above-mentioned grades, cardamom indented for export is evaluated for cleanliness, chemical and physical quality. According to the European Spice Association (ESA) document 'Quality Minima' from 2018, which describes the minimum quality standards for dried herbs and spices, Table 6.3 lists out the physical and chemical quality criteria which applies to cardamom.

In India, Indian standard IS 1907:2023 (BIS, 2023) prescribed minimum physical and quality requirements for trading whole cardamom, seeds and powder. The guidance document from American Spice Trade Association (ASTA) "Clean, Safe Spices" (Table 6.4) prescribed the cleanliness specifications of cardamom (ASTA, 2025). The product should comply with the following physical, chemical and microbiological parameters and conformed products may use standard mark in their retail packaging.

The product should comply with the required physical, chemical and microbiological parameters and conformed products may use standard mark in their retail packaging (Table 6.5). This standard applies to dried or dehydrated fruits of small cardamom as defined in IS 1907:2023, which may be used for direct consumption or as an ingredient in food processing. The standard excludes the product used for industrial processing like volatile oil extraction and/or oleoresin extraction.

Apart from these standards, at international level, the Codex Alimentarius Commission (CAC), an international food standards body established jointly by the Food and Agriculture Organization (FAO) and the World Health Organization (WHO) also developed specific standards for various spices through the Codex Committee on Spices and Culinary Herbs (CCSCH), which is hosted and chaired by India. All 188 member countries and 1 member organization (the EU) of the Codex Alimentarius Commission (CAC) have unanimously agreed to the principle of establishing global standards for spices. Any Codex member country can send delegates to participate in the committee's sessions. These standards are voluntary but serve as international benchmarks for trade, recognized by the World Trade Organization (WTO). In 2024, CCSCH adopted the following standards for small cardamom (FAO and WHO, 2024) whole capsules, seeds and powdered seeds (Table 6.6).

Apart from these chemical characteristics, they have standards for physical characteristics of dried cardamom whole, seed and powdered seed such as number of empty and malformed capsules by count, immature and shrivelled

Table 6.3. Physical and chemical quality criteria for cardamom by ESA (ESA, 2018)

Chemical/physical parameter	Value (%)
Ash (% w/w)	Max. 9.0
Acid insoluble ash (% w/w)	Max. 2.5
Moisture (% w/w)	Max. 12
Volatile oil (% v/w)	Min. 4.0

Table 6.4. Cleanliness specifications of Cardamom (ASTA, 2025)

Chemical/physical parameter	Value (%)
Whole insects dead (No.)	4
Excreta mammalian (mg/lb)	3
Excreta other (mg/lb)	1
Mould (% by weight)	1
Insect defiled/infested (% by weight)	1
Extraneous foreign matter (% by weight)	0.5

Table 6.5. Chemical, physical and microbiological parameters of Cardamom as per IS 1907:2023 (BIS, 2023)

Chemical/physical/ microbiological parameter	Whole cardamom (%)	Cardamom seeds (%)	Cardamom powder (%)
Moisture (% w/w)	Max. 10.0	Max. 10.0	Max. 11.0
Total ash (% w/w)	Max. 8.5	Max. 8.5	Max. 8.0
Acid insoluble ash (% w/w)	-	-	Max. 3.0
Volatile oil (% v/w)	Min. 3.5	Min. 4.0	Min. 3.0
Total aflatoxin (µg/kg)	Max. 30	Max. 30	Max. 30
Aflatoxin B1 (µg/Kg)	Max. 15	Max. 15	Max. 15
Salmonella (cfu/g)	Absent	Absent	Absent
Bacillus cereus (cfu/g)	1×10^4	1×10^4	1×10^4
Sulfite Reducing Clostridia (SRC) (cfu/g)	1×10^3	1×10^3	1×10^3

Table 6.6. Codex standards for small cardamom (FAO and WHO, 2024)

	Whole capsule	Seeds	Seed powder
Chemical/physical parameter	Value (%)		
Ash (% w/w)	Max. 9.5	Max. 9.5	Max. 8.0
Acid insoluble ash (% w/w)	Max. 2.5	Max. 3.0	Max. 3.0
Moisture (% w/w)	Max. 13	Max. 13	Max. 11
Volatile oil (% v/w)	Min. 4.0	Min. 3.5	Min. 3.0

capsules (%), light seeds (%) (in case of cardamom seeds), insect defiled/ infested (%), extraneous matter (%), foreign matter (%), whole dead insects (by count), live insects (by count), mammalian excreta (mg/kg), other excreta (mg/kg) and visible mould (%) (FAO and WHO, 2024).

6.3 Food Safety Standards for Export of Cardamom

All spices should meet the quality standards fixed by any importing countries. In India, the food safety is governed by Food Safety and Standards Authority of India (FSSAI). Likewise, every importing country has their own food safety authority like United States Food and Drug Administration (USFDA), European Food Safety Authority (EFSA), Food Standards Australia New Zealand (FSANZ), Canadian Food Inspection Agency (CFIA), The Food Safety Commission of Japan (FSCJ), Saudi Arabia Food and Drug Authority (SFDA), etc.

The quality and safety standards imposed by the EU on spice imports, especially from India are stringent, mainly to protect the health of its consumer, ensure traceability and promote sustainable farming and to prevent food fraud. EU established EFSA to provide independent analysis of health risks associated with foods and feeds available to European consumers. They shall forward to the European Commission the list of contaminants, microbial toxins, heavy metals, residues and additives used in foodstuffs, as well as other relevant information such as maximum residue levels (MRLs), specific conditions of use and maximum quantities, etc. The exporting countries should adhere to the quality standards of importing countries and non-compliance can lead to rejections, recalls or temporary bans.

Concerning cardamom export, India faced 16 RASFF (Rapid Alert System for Food and Feed) alerts since 2021, most of which were pesticide-related. Farmers use excessive non-label pesticides to combat pests in intensive farming, which resulted in high residue levels, leading to frequent export rejections and bans. A survey conducted between 2008 and 2013 among cardamom planters by Murugan *et al.* (2017) showed that about 26.21 kg a.i/ha of pesticides per year was used to control major pests and diseases in cardamom.

European Commission's EU pesticide database lists around 500 pesticides and their MRL level for cardamom. The range of MRL of different pesticides includes 0.05 mg/kg to 0.5 mg/kg. The level exceeds 0.05 mg/kg in very few pesticides. The exporter should meet out the MRL requirement of those listed pesticides by testing their produce in an accredited testing laboratory accepted by the EFSA and produce the certificate to that effect before exporting (https://ec.europa.eu/food/plant/pesticides/eu-pesticides-db_en). Failing to adhere to the MRL limit may lead to rejection of the shipment. For example, the maximum residue level set by the EU for the pesticide quinalphos is 0.05 mg/kg. According to the EFSA report from 2020, the combined quinalphos content detected in cardamom from India and Guatemala is over 1 mg/kg (Craig *et al.*, 2020). The situation is similar in Japan, where although no specific maximum residue levels are listed for cardamom, around 300 pesticides and their MRLs are listed in the product category 'other spices', which may also apply to cardamom. Among the various countries, the MRLs in the EU and Japan were particularly strict.

The USFDA pesticide database listed 26 pesticides and its MRL level for the index commodity cardamom under herbs and spices group. The Codex Alimentarious Commission has fixed MRL levels for six pesticides for the product cardamom (HS 0775).

The Spices Board advises exporters on MRLs for pesticides and other chemicals from various importing countries for spices and refers them to the official pesticide databases of these countries for accurate and up-to-date information. In the case of the EU, where no MRL has been set for a particular pesticide for a particular spice, a standard MRL of 0.01 mg/kg is generally applied. However, it is essential to verify the database for any spice-specific exceptions. Apart from that, the Spices Board is undertaking several initiatives by conducting campaigns and training programmes in the cardamom growing areas to educate the cultivators on the technologies of integrated pest management (IPM) and good agricultural practices (GAP) in cardamom cultivation. They also encourage growers to use bio-control agents for the management of major pests and diseases to avoid the uncontrolled use of pesticides. Farmers' attitudes towards the indiscriminate use of pesticides have also changed due to export rejections, price fluctuations and damage to natural enemies and cardamom ecosystems. Thanks to these efforts, Indian cardamom exports were rejected less frequently in 2025, with a rejection rate of only 0.5% (www.indianspices.com).

6.4 Conclusion and Outlook

We live in an age where consumers are increasingly aware of the impact that the products they use have on the environment, health and human ethics, and are willing to pay a higher price for clean label and sustainable products. Even governments are gearing their policies towards sustainability. For example, the EU Directive on Corporate Sustainability Due Diligence (CSDDD) requires large EU and non-EU companies operating in the EU to exercise due diligence to prevent and mitigate adverse impacts on human rights and the environment in their operations and value chains. It would be prudent that similar initiatives should be taken by the country from where the spices are grown to comply with the requirements of importing countries like the EU.

India should take the following measures to strengthen its presence in the global cardamom market and meet the growing demand for sustainable products. The implementation of GAP in cardamom cultivation should be done by phasing out the high-risk pesticides and advocating the use of safe botanicals and bio-agents. The infrastructure for postharvest treatment and the certification programme should be made accessible through institutional and political support. By thinking long term, investment should be made for developing climate-resilient varieties to secure the supply chain of future demand of cardamom. By executing this integrated roadmap, India can not only meet but also exceed the stringent quality, traceability and sustainability expectations of Europe and North America, while ensuring that farmers receive premium and stable prices for their certified, residue-free produce.

7

Marketing and Export Prospects of Cardamom

A. Indhushree*

Department of Agricultural Economics, SRM College of Agricultural Sciences, SRM Institute of Science and Technology, Chengalpattu – 603201, India

7.1 Marketing and Trade

From an economic perspective, cardamom is one of the highest-value spices in India's production basket. Although its share in total spice production is small compared to pepper or chilli, its contribution to spice export earnings is disproportionately high because of its premium price in international markets (Spices Board, 2025a). This highlights the importance of cardamom not only as a commercial crop for farmers but also as a source of foreign currency for the national economy. In states such as Kerala and Karnataka, cardamom cultivation plays a stabilising role by diversifying agricultural incomes and reducing dependence on volatile crops such as coffee and pepper, thereby strengthening household resilience. Being a perennial plantation crop requiring intensive labour inputs, particularly during the harvesting season, cardamom plantations support steady demand for rural labour and contribute significantly to regional wage income.

Cardamom farming is highly sensitive to ecological conditions, requiring shaded environments, high rainfall and skilled labour. Despite favourable agro-climatic conditions, cardamom production in India faces challenges due to price volatility, pest infestations, rising input costs and climate change (Murugan *et al.*, 2024). Small cardamom is graded based on certain qualities such as size, weight and colour with higher quality cardamom fetching a higher price in both domestic and export markets. The grades of exportable small cardamom from India are given in Table 7.1.

With globalization, the marketing of cardamom has undergone structural changes. Domestic marketing operates primarily through an auction system regulated by the Spices Board of India, with e-auctions introduced to enhance

*Corresponding author: indhuashree@gmail.com

Table 7.1. Grades of small cardamom from India.

Grade	Size (mm)	Weight (g/l)	Colour
Super bold	8 or above	450	Matured green
Extra bold	7 or above	435	Matured green
Bold	6.5 or above	415	Light green
Small	5.5 to 6.5	385	Light green

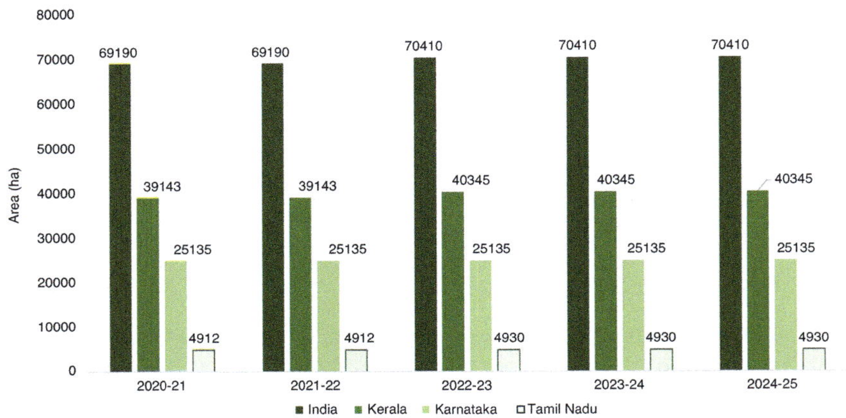

Fig. 7.1. Area (hectares) of small cardamom in major cardamom producing states of India from 2020–2021 to 2024–2025.

transparency. On the export front, India's cardamom trade is concentrated in the Middle East, particularly Saudi Arabia, UAE, Kuwait and Qatar, which together absorb more than 60% of exports (Indhushree and Kuruvila, 2019; Murugan *et al.*, 2024).

7.2 Production and Supply Trends of Cardamom

India ranks second to Guatemala in small cardamom production. In 2023, Guatemala contributed 43.43% to global cardamom production (Tridge, 2025), while India's contribution was 24.64%. The production of small cardamom in Guatemala has increased by 127.43% during the period 2019 to 2023. India's production of small cardamom has shown both resilience and volatility in recent years. The area under cultivation has remained constant at around 70,000 ha (Fig. 7.1), whereas production increased from 22,520 Metric Tonnes (MT) in 2020–2021 to 25,230 MT in 2023–2024 before declining to 20,696 MT in 2024–2025 as seen in Fig. 7.2 (Spices Board, 2025b). Kerala continues to dominate in both area (Fig. 7.1) and production (Fig. 7.2), followed by Tamil Nadu and Karnataka. Climatic conditions, pest outbreaks and replanting cycles are the major factors influencing the cardamom productivity.

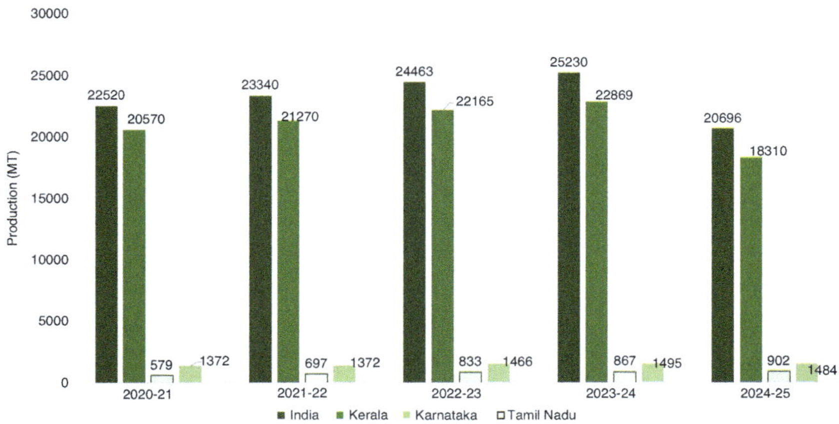

Fig. 7.2. Production (MT) of small cardamom in major cardamom producing states of India from 2020–2021 to 2024–2025.

7.3 Domestic Marketing System of Cardamom

The marketing of small cardamom in India has traditionally been carried out through regulated auction systems. The states of Kerala, Karnataka and Tamil Nadu are the major centres where cardamom auctions are conducted under the supervision of the Spices Board of India. The introduction of electronic auctions in 2007 marked a major reform by enhancing transparency, reducing collusion and improving efficiency in price discovery. Farmers bring their produce to the auction centres, where registered traders and exporters bid competitively. Registered auctioneers play a major role in conducting the e-auctions. The quality of cardamom, based on size, colour and aroma, plays a significant role in price determination.

Apart from the auction system, cardamom is marketed through cooperative societies, private traders and direct sales to exporters. The Spices Board has also promoted branding initiatives such as Geographical Indications (GI) for Malabar and Coorg cardamom, which can help improve market positioning. Re-pooling was a major problem in the auction system, where dealers purchase cardamom in the auction and pool it and sell in another auction for good profit. This led to manipulation and irregularity in the entire auction system. In 2024, Spices Board of India imposed restriction on re-pooling of cardamom in the auction centre. The licensed auction companies were directed to enforce the rule and ensure that re-pooling should not exceed more than 25% of total quantity of cardamom put in auction (Vellaram, 2024).

7.4 Export Performance of Indian Cardamom

India's export of small cardamom has experienced significant fluctuations over the decades. Historically, India was the world's leading exporter, but

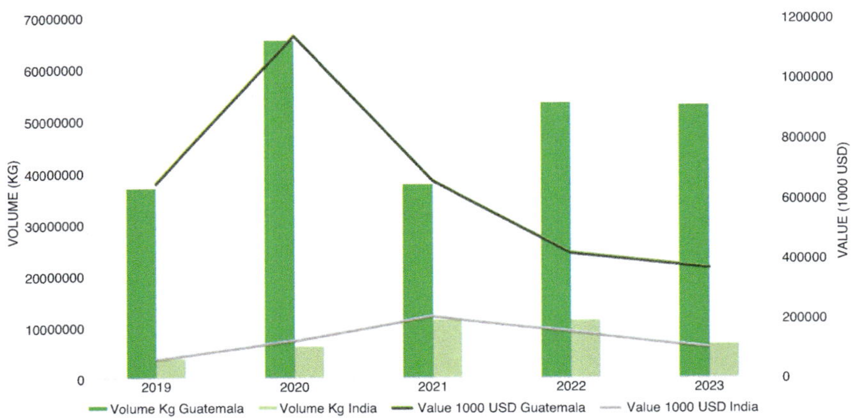

Fig. 7.3. Volume and value of cardamom (HS code 090830: Cardamom) export from Guatemala and India from 2019 to 2023.

Guatemala has emerged as the dominant supplier in recent years. In 2023, Guatemala accounted for nearly 53.28% of global cardamom export in terms of value (Fig. 7.3). But, Indian cardamom maintains a strong reputation for quality, aroma and flavour, fetching premium prices in international markets, particularly in the Gulf countries. Indian cardamom is known for its stronger flavour and longer-lasting green colour, giving it a unique edge in both domestic and global markets. In addition, India benefits from a well-regulated system overseen by the Spices Board of India, which operates an e-auction platform for wholesalers and exporters, thereby reducing speculation and ensuring greater transparency. Most importantly, India consumes about 98% of its total production, making its export prices highly sensitive to domestic market conditions. In contrast, Guatemala exports nearly 100% of its production, making it more export-driven than India.

According to Spices Board data, India exported more than 147,000 tonnes of small cardamom in 2024–2025, which marked an all-time record. Exports were also very high in 2021–2022, at nearly 120,000 tonnes, though there was a noticeable decline to around 80,000 tonnes in 2022–2023 before recovering steadily in subsequent years (Spices Board, 2025c). This shows that while there are fluctuations, Indian small cardamom exports maintain a far stronger international presence.

The Gulf countries, especially the United Arab Emirates and Saudi Arabia, dominate the demand for small cardamom and consistently remain the largest buyers (Fig. 7.4). The USA also continues to be a significant market, while countries such as Kuwait, Qatar and Oman collectively reinforce the importance of the Gulf region as the backbone of India's small cardamom exports. Bangladesh showed an exceptional spike in demand in 2021–2022, but its imports fell sharply afterwards, suggesting that its trade relationship with India is less stable. Traditional markets such as the United Kingdom, Canada

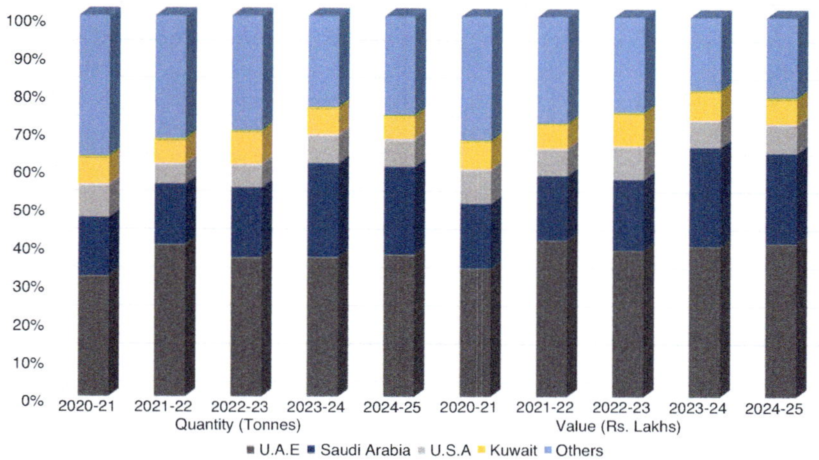

Fig. 7.4. Percentage share of different countries in quantity and value of small cardamom exported from India during 2020–2021 to 2024–2025.

and Iran provide steady support, but their volumes are secondary compared to the Gulf. This dependence on a few markets makes exports vulnerable to geopolitics, policy changes, price sensitivity, or demand shifts in that region.

Cardamom exports from India have undergone major changes over the past few decades. Guatemala, which produced only about 800 tonnes of cardamom in 1970, increased its production nearly 20 times within three decades (FAOSTAT, 2024; ITC Trade Map, 2023). This rapid growth, coupled with rising domestic demand in India, led to a steady decline in India's export share. In the 1970s, India exported nearly half of its production and held about 50% of global exports. However, India's share declined systematically from 48.5% in 1989–1990 to 8.7% in 2017–2018 (Murugan *et al.*, 2024; Indhushree and Kuruvila, 2019). The proportion of domestic production exported also declined, reaching as low as 10.1% in the early 2000s, though it later revived to 22.3% by 2017–2018. Interestingly, India's export performance tended to improve only when Guatemala faced crop failures due to pests or adverse weather conditions (George and Cherian, 2017).

Another significant change has been noticed in the composition of exports. While unprocessed cardamom dominated exports earlier, the share of processed and value-added products has increased from 5% in 2000 to nearly 27% in 2018. This shift reflects better profit opportunities in processed products and higher demand for value addition (DGCI&S, 2019; Murugan *et al.*, 2024), which also generates more employment along the value chain.

India's competitiveness in global markets declined sharply from the early 1990s to 2010, largely because of Guatemala's rapid production growth and India's high-cost production system (Thomas *et al.*, 2019). However, Indian cardamom consistently commanded higher unit export values (US$18–25/kg) compared to Guatemala's US$8–12/kg, owing to its superior aroma and quality (ITC Trade Map, 2023). After 2010, India's competitiveness improved slightly, partly because of institutional and production challenges in Guatemala.

To strengthen its export competitiveness, India needs to lower its production costs by rationalising the use of inputs, adopting crop management technologies, investing in research and introducing mechanisation in labour-intensive operations. Such measures can ensure that Indian cardamom not only maintains its premium image but also competes effectively with Guatemala and other emerging producers.

7.5 Export Value Chain of Indian Cardamom

The export value chain for cardamom begins at the farm and passes through multiple actors before the product reaches overseas buyers. Primary producers, mostly small and marginal farmers in Kerala, Karnataka and Tamil Nadu, sell to local traders or cooperative societies. These aggregators supply to licensed auction centres or directly to wholesale traders and exporters, who perform cleaning, grading, value-addition, packing and export documentation. Exporters play a crucial role in linking domestic production with international markets: they decide on product grade, arrange quality tests and phytosanitary clearances, negotiate contracts with overseas buyers and manage logistics and finance for export shipments (Indhushree and Kuruvila, 2019; Thomas *et al.*, 2019).

The Spices Board of India acts as the nodal government agency supporting nearly every stage of the export chain. The Board runs promotional activities, ensures quality through its accredited testing laboratories, facilitates e-auctions (to improve transparency in primary marketing), supports capacity building of farmer groups and links exporters with overseas buyers through trade fairs and buyer–seller meets. The Board also provides technical guidance on grading, packaging and compliance with importing countries' food safety requirements. The introduction of e-auctions, licensing rules and targeted programmes has improved price discovery and reduced market distortions in the primary market to some extent.

Compliance with customs and legal regulations is another essential link in the chain. Exporters must clear shipments through customs, submit export documentation to the Directorate General of Foreign Trade and the Export-Import (DGCI&S) database, and ensure that consignments meet the maximum residue limits (MRLs) and sanitary and phytosanitary (SPS) requirements of importing countries. Failure to meet MRLs or other SPS norms can result in shipment rejections, costly recalls, or temporary market closures (Case: Saudi Arabia's restrictions on Indian cardamom due to pesticide residues) (Murugan *et al.*, 2024; European Commission, 2025). To reduce such risks, exporters arrange pre-export residue testing, provide traceability documentation and often work with certified labs and accredited testing agencies.

7.6 Composition of India's Cardamom Exports

The structure of India's cardamom (including large and small cardamoms) exports during 2017–2025 reveals a consistent dominance of whole cardamom over crushed and ground forms (Fig. 7.5). Whole cardamom accounted for between 85 and 95% of the total export basket throughout this period, highlighting the sector's continued reliance on bulk, unprocessed shipments. In contrast, crushed and ground cardamom contributed only 4–14%, reflecting the limited extent of value addition in India's export portfolio.

Export volumes of whole cardamom have shown sharp year-to-year fluctuations. Exports fell steeply from 6.48 million kilograms in 2017–2018 to just over 3.1 million kilograms in 2019–2020, before rebounding strongly to 7.16 million kilograms in 2020–2021 and peaking at more than 12 million kilograms in 2021–2022 (Ministry of Commerce, 2025). This peak year coincided with crop constraints in Guatemala, which temporarily improved India's competitiveness in global markets. However, volumes subsequently declined again, stabilising in the range of 7–7.6 million kilograms during 2023–2024 and 2024–2025. These swings reflect the biological and climatic sensitivity of cardamom production, coupled with competitive pressures in the global spice trade.

The contribution of crushed and ground cardamom was relatively higher in 2018–2019 and 2019–2020, when it accounted for about 14% of total exports. Since then, the share has declined to below 7%. This suggests that exporters remain cautious about investing in processing and value-added forms, despite their potential for higher unit values and stronger branding in international markets. The consistently low share of processed exports underscores a structural weakness in India's cardamom trade and highlights a clear opportunity for

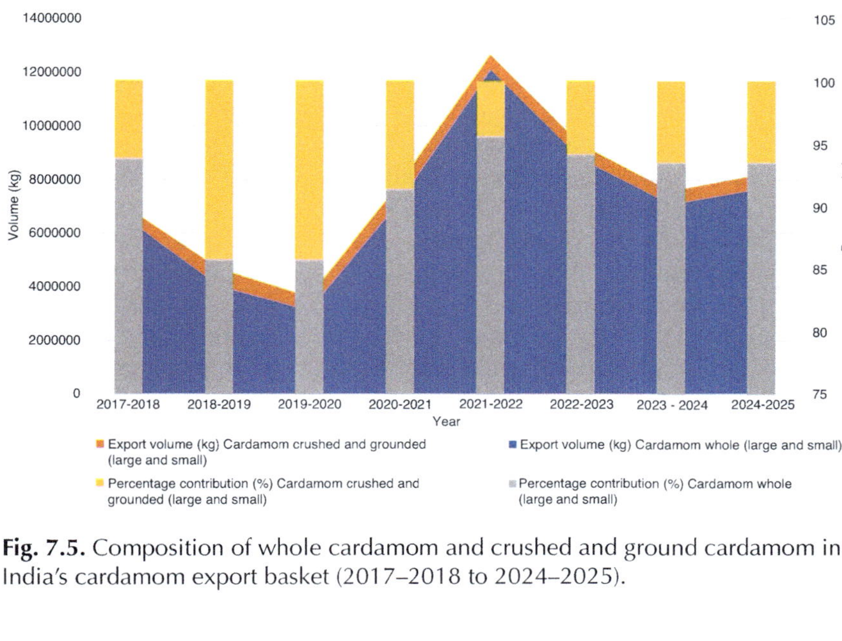

Fig. 7.5. Composition of whole cardamom and crushed and ground cardamom in India's cardamom export basket (2017–2018 to 2024–2025).

diversification through product innovation, branding and promotion of value-added derivatives. Expanding the share of crushed, ground and other value-added forms could help India move up the value chain, stabilise export earnings and reduce vulnerability to fluctuations in whole cardamom production.

7.7 Trend and Opportunities in the Global Green Cardamom Market

The demand for organic cardamom has risen sharply in recent years, reflecting global consumer preferences for natural, sustainable and health-oriented products. Organic cardamom appeals particularly to health-conscious buyers in Europe and North America, where clean-label and chemical-free foods command a premium. Beyond its culinary appeal, cardamom's medicinal qualities have strengthened its presence in herbal medicines, nutraceuticals and wellness supplements. This dual positioning as both a premium spice and a natural health product has expanded its global market potential.

Changing dietary patterns and the globalization of food culture have also boosted cardamom consumption. Once confined mainly to Indian, Middle Eastern and Scandinavian cuisines, green cardamom is increasingly used in Western markets in beverages like chai lattes and craft cocktails, as well as in gourmet desserts and chocolates. Rising disposable incomes in emerging economies have further broadened demand, as consumers show growing interest in high-value spices and exotic flavours. Food companies in the West are capitalising on this trend by incorporating cardamom into innovative product lines that cater to adventurous palates, thereby expanding the spice's reach into mainstream retail and food service sectors (Grand view research, 2025).

Government initiatives in producing countries have played a pivotal role in shaping this growth. In India, the Spices Board, under the Ministry of Commerce and Industry, has supported cardamom producers through subsidies for irrigation, training programmes and investment in disease-resistant and high-yielding varieties. These measures have enhanced both productivity and quality, ensuring greater competitiveness in export markets. Similar efforts in Guatemala, the world's leading exporter, have emphasized sustainability and certification, allowing producer cooperatives to leverage organic and fair-trade branding to meet the demands of high-value markets in Europe and the USA. Conventionally grown cardamom continues to account for the bulk of production; however, the organic segment is expanding at nearly 7% annually, driven by consumer willingness to pay a premium for environmentally sustainable and chemical-free products.

Rising consumer demand for organic and exotic flavours, combined with innovations in processing, branding and certification, offers significant opportunities for producers and exporters. By leveraging government support, investing in postharvest value addition and aligning with global health and sustainability trends, India can sustain as the world's leading supplier of this high-value spice.

7.8 Government Initiatives in Cardamom Market

The Government of India has put in place several schemes and institutional supports to promote spice exports and strengthen value chains. One important instrument is the Market Development Assistance (MDA), which provides financial support to exporters (especially small exporters) and export promotion councils to participate in trade fairs, buyer–seller meets, market studies and branding initiatives. The MDA scheme aims to improve market linkages and reduce marketing costs for exporters (Ministry of Commerce and Industry, 2025; Spices Board, 2025d).

Beyond MDA, the Spices Board runs multiple promotional and development programmes. These include export promotion missions, laboratory and quality assurance support, e-auction platforms to improve farmer price realisation, training and capacity building for processors and farmer groups and targeted schemes (e.g. SPICED) to expand area and improve productivity and postharvest handling as per Spices Board programme pages and SPICED guidelines (Spices Board, 2025e). Such measures are designed to increase the exportable surplus, improve quality compliance and encourage value addition.

Geographical Indication tagging has been used to protect and brand region-specific cardamom types. The Spices Board of India has registered Alleppey Green cardamom and Coorg Green cardamom for GI tag (Spices Board, 2025f). GIs signal origin-specific quality and can help fetch premium prices in niche markets when supported by marketing and traceability. Quality control measures have become central to sustaining market access. The Spices Board supports accredited testing laboratories and works with international partners to help exporters comply with importing countries' MRLs and SPS requirements. India has also been piloting blockchain and QR-code-based traceability projects (with UNDP's Accelerator Lab, GS1 and others) to provide immutable farm-to-fork traceability, which is an intervention designed to accelerate compliance, reduce transaction costs and increase buyer confidence in Indian spices (Spices Board, 2025a). These traceability pilots are important especially for premium and regulated markets where provenance, pesticide history and traceability matter.

7.9 Challenges and Opportunities in Cardamom Marketing and Export

India's cardamom exports are constrained by a mix of production volatility, global competition and regulatory pressures. Being highly sensitive to climate, pests and replanting cycles, cardamom output fluctuates sharply from year to year, which results in unstable auction prices and uncertain returns for exporters. This volatility has been compounded by the massive expansion of Guatemalan production, which has overtaken India in global supply.

Quality concerns add another layer of difficulty. Importing countries such as Saudi Arabia and the EU enforce strict maximum residue limits (MRLs) and sanitary and phytosanitary (SPS) standards. Indian shipments have occasionally faced rejections due to pesticide residues, highlighting the high costs of compliance through safer pest management, residue testing and certification. At the same time, inadequate branding and product differentiation weaken India's competitive edge. While GIs like Alleppey Green and Coorg cardamom exist, exporters still rely heavily on unprocessed bulk shipments, competing largely on price rather than quality or traceability.

Other structural challenges further constrain growth. For example, exports of cardamom remain heavily concentrated in Gulf markets, leaving India vulnerable to regulatory or demand shifts in just a few countries. Limited diversification into EU, North America, or East Asia is hindered by stricter SPS and technical barriers. Postharvest weaknesses, such as poor grading, packaging and storage infrastructure, also reduce shelf life and market value. Under World Trade Organisation (WTO) trade rules, declining tariffs have increased exposure to cheaper imports, while a rise in non-tariff measures such as SPS regulations has raised compliance costs for exporters, particularly smallholders (World Trade Organization, 2022). Addressing these challenges through improved infrastructure, branding and diversification is essential for India to sustain and expand its position in global cardamom trade.

Despite current challenges, India's cardamom sector holds significant opportunities. Global demand for natural flavours, nutraceuticals and organic spices is on the rise, with consumers in developed markets increasingly seeking plant-based and sustainably sourced products. Cardamom, with its strong aromatic and health attributes, is well positioned to benefit from this trend. Certified organic cardamom from sustainable production fetches top prices in niche markets, particularly in Europe and North America (Thomas *et al.*, 2019).

Expanding beyond the Middle East, markets in the EU, USA and East Asia offer scope for higher-value exports, provided Indian exporters invest in compliance with strict standards on residues, hygiene and traceability. Greater emphasis on branding through GI tags and government-supported market development programmes can help India building a stronger presence in these regions. At the same time, value addition offers another pathway for growth. Exports of cardamom powders, essential oils, oleoresins and premium blends are increasing, and further investment in processing, packaging and product innovation can contribute to both diversify markets and create local employment, while improving farmers' returns.

Technology and institutional support will also be critical. As digital tools, namely blockchain and QR code-based traceability, already being piloted for spices, they could help India gain buyer confidence in premium markets. Similarly, converting select producing regions to certified organic systems, backed by farmer producer organizations (FPOs) and non-governmental organizations (NGOs), can strengthen India's market position. Policy initiatives in Research and Development (R&D), mechanization and integrated pest management will not only lower costs but also ensure compliance with international standards. With coordinated efforts in innovation, branding and institutional support, India can reinforce its identity as a producer of premium-quality cardamom and expand its role in the global spice economy.

7.10 Conclusion

The export value chain of Indian cardamom reflects both strengths and vulnerabilities. Premium quality continues to sustain demand, yet high volatility, strict regulatory compliance and weak branding, limit growth. Opportunities in organic certification, value addition and traceability offer new directions, particularly in EU, USA and East Asian markets. Achieving these gains requires policy support, investments in research and postharvest infrastructure and stronger producer organizations. With coordinated reforms, India can consolidate its position as a benchmark for sustainable and resilient spice exports. Despite India's reputation for premium-quality cardamom, the sector faces persistent challenges. Price volatility, often driven by climatic variability and production shocks, undermines farmer incomes and market stability. Stringent pesticide residue limits imposed by importing countries present additional hurdles, while weak branding strategies, infrastructural deficiencies in grading, storage and packaging, and dependence on a few Gulf markets restrict the sector's growth potential.

Nevertheless, significant opportunities remain for enhancing India's position in global cardamom trade. Value addition through the development of powders, essential oils and blends can increase profitability and employment opportunities. Organic cultivation and certification, backed by consumer willingness to pay a premium, offer a niche advantage in Western markets. Emerging technologies such as blockchain and QR-based traceability systems can strengthen buyer confidence and open access to premium segments.

To realize these opportunities, India must invest strategically in research and development, particularly in pest-resistant varieties and cost-reducing technologies, while simultaneously building robust postharvest infrastructure. StrengtheningFPOs will also be critical to aggregating supply, improving bargaining power and facilitating direct export linkages. With coordinated efforts in innovation, institutional support and policy reform along with farmer capacity building and targeted export promotion, India can reinforce its legacy as a producer of high-quality cardamom and secure a stronger and more sustainable role in the global spice economy.

8

Future Prospects in Cardamom Research and Development

H.J. Akshitha* and R. Sivaranjani

ICAR-Indian Institute of Spices Research, Kozhikode, Kerala

8.1 Research and Development Goals

Research and development efforts in India are increasingly focused on developing high-yielding cardamom varieties with improved resistance to major biotic stresses such as viral diseases, pests and fungal pathogens, as well as abiotic stresses like drought, temperature fluctuations and soil nutrient imbalances. Parallel emphasis is being placed on promoting sustainable agronomic practices, refining organic production packages and strengthening integrated pest and disease management (IPDM) strategies to enhance productivity while reducing environmental impact. These initiatives are further supported by advancements in precision farming tools, improved postharvest handling techniques and quality-enhancement protocols aimed at maintaining the spice's export competitiveness. Economically, the cardamom sector continues to show promising growth, with the global market valued at US$890.5 million in 2024 and projected to reach approximately US$1655.7 million by 2035, growing at a compound annual growth rate of about 5.8% between 2025 and 2035. This expanding market underscores the importance of sustained R&D efforts to meet rising global demand, ensure supply stability and enhance the value-addition potential of cardamom-based products.

8.2 Challenges in Cardamom Cultivation

Over the years, the area under cardamom cultivation in India has remained relatively stable, even though both production and export volumes have shown an upward trend. Despite the increase in export quantities, the export value of cardamom has been declining (Gangadharan *et al.*, 2025). The trends in cultivation area, production and productivity are influenced by several challenges

*Corresponding author: akshi.hosahalli10@gmail.com

faced by cardamom growers. Major challenges include (Nadukeri *et al.*, 2018; Jegadeesan and Miyazaki, 2025):

1. **Changing climatic conditions**: Erratic rainfall, floods, severe droughts and high temperatures adversely affect crop growth and yield.
2. **Pest and disease infestations**: The outbreak of pests and diseases such as thrips infestation, capsule and shoot borer, katte disease, leaf blight and rhizome rot, which can significantly reduce productivity.
3. **Limited availability of improved planting material**: The perennial nature of crop and limited access to high-yielding and disease-resistant varieties hinders production.
4. **High input costs and labour-intensive cultivation**: Cardamom farming requires substantial investment in fertilizers, pesticides and labour, making it economically challenging for small-scale growers.
5. **Market volatility**: The uncertainty created by local and international demand and related price fluctuation affects farmers' profitability.
6. **Strict import regulations**: International trade restrictions and quality standards can limit market access and export potential.

8.3 Future Prospects

- **Development of climate-resilient, high-yielding and stress-tolerant varieties**: Future breeding programmes must prioritise the development of cardamom varieties capable of withstanding a wide range of biotic and abiotic stresses. This includes resistance to major pests (thrips, borers) and diseases (rhizome rot, viral infections), as well as tolerance to drought, high temperatures, erratic rainfall and soil nutrient imbalances. Integrating modern approaches such as molecular breeding, marker-assisted selection, genomic selection and biotechnology will accelerate the release of high-yielding, climate-smart cultivars that ensure stable productivity under changing environmental conditions.
- **Advancement of sustainable and smart farming technologies**: The adoption of smart agriculture tools will be essential for improving efficiency and resource-use optimisation in cardamom plantations. Sensor-based irrigation systems, automated microclimate monitoring, precision nutrient management and digital farm advisory platforms can significantly enhance productivity while minimising input use. The development of location-specific cultivation packages, mechanisation of key operations (such as shade regulation and weed management) and the integration of IoT and AI-driven decision-support systems will promote sustainable and economically viable cultivation.
- **Promotion of organic and integrated pest and disease management practices**: Future research must focus on eco-friendly and integrated management strategies that reduce dependence on chemical pesticides.

This includes developing bio-control agents, microbial consortia, botanical extracts, pheromone traps and disease-forecasting models. Strengthening organic nutrient management using bio-fertilisers, composts and microbial inoculants will support soil health and long-term sustainability. A robust IPDM framework will help maintain environmental safety, improve cardamom quality and enhance export competitiveness.

- **Development of synchronous and reduced harvest varieties to lower labour dependence**: Cardamom harvesting is highly labour-intensive due to the continuous and staggered emergence of panicles and capsules. Future breeding and crop improvement programmes should emphasize the development of synchronous flowering and fruiting varieties or cultivars that require fewer harvest cycles, thereby reducing labour requirements and production costs. Such innovations will be especially beneficial in regions experiencing acute labour shortages and rising wage rates.
- **Improvement of postharvest technologies, value addition and industrial extraction**: Enhancing postharvest handling is critical for maintaining the distinctive flavour and colour of cardamom. Research should focus on developing energy-efficient dryers, improved curing techniques, scientific grading and packaging systems and storage technologies that preserve quality. There is significant potential for product diversification, including the extraction of essential oils, oleoresins and flavour compounds for use in the food, pharmaceutical, cosmetic and nutraceutical industries. Establishing small-scale processing units and value-addition enterprises can boost farmer income and create rural employment.
- **Strengthening market intelligence and digital marketing platforms**: Price volatility and market uncertainty are major challenges in the cardamom sector. A robust market intelligence system integrating real-time price tracking, demand forecasting, export trends and early warning systems can help farmers make informed decisions regarding harvest, storage and sales. Digital platforms and mobile-based advisory services can improve farmer access to markets, reduce dependence on middlemen and minimise losses. Research should also explore blockchain-based traceability, e-commerce integration and branding strategies to enhance global market presence.

8.4 Conclusion

The future of cardamom R&D is not only rich with opportunities but also demands a multi-disciplinary, systems-based approach. Key focus areas should include climate resilience, advanced breeding, integrated pest management, postharvest technology, value addition, biodiversity conservation, market diversification and socio-economic research. In order to realise these prospects and sustainably increase cardamom production and profitability, coordinated efforts by research institutions, government agencies, farmers and the private sector are essential.

References

Abdullah, A., Asghar, A., Butt, M.S., Shahid, M. and Huang, Q. (2017) Evaluating the antimicrobial potential of green cardamom essential oil focusing on quorum sensing inhibition of *Chromobacterium violaceum*. *Journal of Food Science and Technology* 54, 2306–2315. DOI: 10.1007/s13197-017-2668-7.

Abdullah, A., Asghar, A., Algburi, A., Huang, Q., Ahmad, T. *et al.* (2021) Anti-biofilm potential of *Elettaria cardamomum* essential oil against *Escherichia coli* O157:H7 and *Salmonella typhimurium* JSG 1748. *Frontiers in Microbiology* 12, 620227. DOI: 10.3389/fmicb.2021.620227.

Aboubakr, M. and Abdelazem, A.M. (2016) Hepatoprotective effect of aqueous extract of cardamom against gentamicin-induced hepatic damage in rats. *International Journal of Basic and Applied Sciences* 5(1), 1–4.

Abraham, P. and Tulasidas, G. (1958) South Indian cardamoms and their agricultural value. *ICAR Technical Bulletin* 79, 1–27.

Adireddy, R.G., Anapalli, S.S., Reddy, K.N. and Gowda, P.H. (2025) Quantifying the interactive effects of temperature, CO_2, and irrigation levels on peanut yield and quality. *Plant Stress* 17, 100959. DOI: 10.1016/j.stress.2025.100959.

Aghasi, M., Ghazi-Zahedi, S., Koohdani, F., Siassi, F., Nasli-Esfahani, E. *et al.* (2018) The effects of green cardamom supplementation on blood glucose, lipid profile, oxidative stress, sirtuin-1 and irisin in type 2 diabetic patients: A randomized placebo-controlled clinical trial. *BMC Complementary Medicine and Therapies* 18(18), 1–9. DOI: 10.1186/s12906-017-2068-6.

Ahamad, J. (2024) Anticancer potential of cardamom (*Elettaria cardamomum*) essential oil against TNBC, glioma and kidney cancer *in vitro*. *Journal of Angiotherapy* 8(9), 1–7. DOI: 10.25163/angiotherapy.889859.

Akhila, P.R., Radhakrishnan, V.V., Pradipkumar, K. and Mohanan, K.V. (2017) Assessment of variability and performance of new landraces of small cardamom (*Elettaria cardamomum* Maton). *Journal of Plantation Crops* 45, 43–48.

Akshitha, H.J., Peeran, M.F., Sharon, A. and Balaji, R.M. (2021) Evaluation of small cardamom (*Elettaria cardamomum* Maton) hybrids for yield and rhizome rot resistance. In: *Proceedings of the International Symposium on Spices and Aromatic Crops (SYMSAC X) – Spices as Flavours, Fragrances and Functional Foods*, Indian Society for Spices, February 9–12, 2021.

Akshitha, H.J., Shivakumar, M.S. and Ankegowda, S.J. (2024) Phenological growth stages of cardamom (*Elettaria cardamomum* Maton): Detailed identification and description using the extended BBCH scale. *Annals of Applied Biology* 184(3), 391–399. DOI: 10.1111/aab.12895.

Alagupalamuthirsolai, M., Ankegowda, S.J., Murugan, M., Sivaranjani, R., Rajkumar, B. *et al.* (2019) Influence of Light Intensity on Photosynthesis, Capsule Yield, Essential

Oil and Insect Pest Incidence of Small Cardamom (*Elettaria cardamomum* (L.) Maton). *Journal of Essential Oil Bearing Plants* 22(5), 1172–1181.

Alagupalamuthirsolai, M., Murugan, M., Ashokkumar, K., Ankegowda, S.J., Vellaikumar, S. *et al.* (2024) Bio-active essential oil components in small cardamom as influenced by weather elements. *Agricultural Mechanization in Asia, Africa and Latin America* 55(4), 17549–17560.

Alam, A., Jawaid, T. and Alam, P. (2021) *In vitro* antioxidant and anti-inflammatory activities of green cardamom essential oil and in silico molecular docking of its major bioactives. *Journal of Taibah University for Science* 15(1), 757–768. DOI: 10.1080/16583655.2021.2002550.

Al-Ismail; K.M., Mehyar, G., Al-Khatib, H.S. and Al-Dabbas, M. (2015) Effect of micro encapsulation of cardamom's essential oil in gum Arabic and whey protein isolate using spray drying on its stability during storage. *Quality Assurance and Safety of Crops & Foods* 7(5), 613–620. DOI: 10.3920/QAS2014.0422.

Alkhalifah, E.A., Alobaid, A.A., Almajed, M.A., Alomair, M.K., Alabduladheem, L.S. *et al.* (2022) Cardamom extract alleviates the oxidative stress, inflammation and apoptosis induced during acetaminophen-induced hepatic toxicity via modulating Nrf2/HO-1/NQO-1 pathway. *Current Issues in Molecular Biology* 44(11), 5390–5404. DOI: 10.3390/cimb44110365.

Alniemi, A.N. (2010) Production of cardamom plants (*Elettaria cardamomum* Maton) through one-step regeneration of embryos stimulated in vitro. *Mesopotamia Journal of Agriculture* 38, 188–197. Available at: https://doaj.org/article/cf3a5382610243c29f22f1db5071acfe

Andriza, Y., Rachmawati, A., Diastuti, H., Agustian, E. and Sulaswatty, A. (2022) The effect of ultrasonic pre-treatment on yield of hydrodistillation cardamom (*Amomum cardamomum*) seeds essential oil. *AIP Conference Proceedings* 2493(1), 050010. DOI: 10.1063/5.0110916.

Anjali, N., Dharan, S.S., Nadiya, F. and Sabu, K.K. (2015) Development of EST-SSR markers to assess genetic diversity in *Elettaria cardamomum* Maton. *International Journal of Applied Sciences and Biotechnology* 3(2), 264–269.

Anjali, N., Nadiya, F., Shefeek, S. and Sabu, K.K. (2016a) Genome size estimation using flow cytometry in *Elettaria Cardamomum* Maton. *Indian Forester* 142(9), 878–881.

Anjali, N., Ganga, K.M., Nadiya, F., Shefeek, S. and Sabu, K.K. (2016b) Intraspecific variations in cardamom (*Elettaria cardamomum* Maton): Assessment of genomic diversity by flow cytometry, cytological studies and ISSR analysis. *Springer Plus* 5, 1560. DOI: 10.1186/s40064-016-3226-x.

Ankegowda, S.J. and Krishnamurthy, K.S. (1998) Effect of moisture stress on growth of cardamom (*Elettaria cardamomum* Maton). In: Sadanandan, A.K., Krishnamurthy, K.S., Kandamman, K. and Karikanthianath, V.S. (eds) *Water and Nutrient Management for Sustainable Production and Quality of Spices*. Indian Society for Spices, Calicut, pp. 158–161.

Ankegowda, S.J. and Krishnamurthy, K.S. (2008) Evaluation of small cardamom accessions for moisture stress. *Journal of Spices and Aromatic Crops* 17(2), 172–176.

Ankegowda, S.J., Bhat, A.I., Jayasree, E., Leela, N.K., Kumar, S.C.M. *et al.* (2022a) Cardamom. Extension pamphlet 2022/01. ICAR-Indian Institute of Spices Research, Kozhikode, Kerala, pp. 1–32.

Ankegowda, S.J., Alagupalamuthirsolai, M., Sivaranjani, R., Mohammed Faisal, P., Krishnamurthy, K.S. *et al.* (2022b) Variability in essential oil constituents, antioxidant activities and yield of elite small cardamom lines (*Elettaria cardamomum* Maton) under moisture deficit stress. *Journal of Essential Oil Bearing Plants* 25(5), 1062–1076. DOI: 10.1080/0972060X.2022.2125351.

Anonymous (2003) In: *The Ayurvedic Formulary of India*. Government of India, Ministry of Health and Family Welfare, Department of Ayurveda, Yoga & Naturopathy, Unani, Siddha and Homoeopathy, New Delhi.

Anonymous (2011) Caraka*: Caraka saṃhitā* (Reprint). Chaukhambha Surabharati Prakashan, Varanasi.

Anonymous (2022) *Elettaria cardamomum* (L.) Maton. *The Plant List*. Available at: http://www.theplantlist.org/tpl1.1/record/kew-243056

Asbaghi, O., Eslampour, E., Reiner, Ž., Badehnoosh, B., Kolahdooz, F. *et al.* (2020) Effect of green cardamom on lipoproteins, glycemic control and anthropometric parameters: A meta-analysis of randomized clinical trials. *Clinical Nutrition ESPEN* 37, 24–33. DOI: 10.1016/j.clnesp.2020.03.015.

Ashokkumar, K., Dhanya, M.K., Murugan, M., Haritha, K., Nimisha, M. *et al.* (2020a) Assessment of variation in volatile oil content in cardamom [*Elettaria cardamomum* (L.) Maton]. *Electronic Journal of Plant Breeding* 11(4), 1007–1014. DOI: 10.37992/2020.1104.164.

Ashokkumar, K., Pandian, A., Murugan, M., Dhanya, M.K., Sathyan, T. *et al.* (2020b) Profiling bioactive flavonoids and carotenoids in select South Indian spices and nuts. *Natural Product Research* 34(9), 1306–1310. DOI: 10.1080/14786419.2018.1557179.

Ashokkumar, K., Murugan, M., Dhanya, M., Raj, S. and Kamaraj, D. (2020c) Phytochemical variations among four distinct varieties of Indian cardamom *Elettaria cardamomum* (L.) Maton. *Natural Product Research* 34, 1919–1922. DOI: 10.1080/14786419.2018.1561687.

Ashokkumar, K., Vellaikumar, S., Murugan, M., Dhanya, M.K., Ariharasutharsan, G. *et al.* (2021) Essential oil profile diversity in cardamom accessions from Southern India. *Frontiers in Sustainable Food Systems* 5, 639619. DOI: 10.3389/fsufs.2021.639619.

Ashour, M.N., Elias, T.R., Rasheed, W.I., El-Toukhya, S.E., Youness, E.R. *et al.* (2024) Neuroprotective effect of cardamom essential oil encapsulated nanoparticles and swimming in aluminum-induced Alzheimer's disease in rats. *Egyptian Journal of Chemistry* 67(1).

ASTA (2025) Available at: https://astaspice.org/resources/asta-cleanliness-specifications

Auti, S.T. and Kulkarni, Y.A. (2019) Neuroprotective effect of cardamom oil against aluminium-induced neurotoxicity in rats. *Frontiers in Neurology* 10, 399. DOI: 10.3389/fneur.2019.00399.

Babu, K.N., Ravindran, P.N. and Peter, K.V. (1997) *Protocols of micropropagation of spices and aromatic crops*. Indian Institute of Spices Research Calicut, Kerala, India.

Babu, K.N., Geetha, S.P., Minoo, D., Ravindran, P.N. and Peter, K.V. (1999) *In vitro* conservation of cardamom (*Elettaria cardamomum* Maton) germplasm. *Plant Genetic Resources Newsletter* 119, 41–45.

Bano, S., Majumder, A., Srivastava, A. and Nayak, K.B. (2024) Deciphering the potentials of cardamom in cancer prevention and therapy: From kitchen to clinic. *Biomolecules* 14(9), 1166. DOI: 10.3390/biom14091166.

Bavappa, K.V.A. (1986) Research at CPCRI. In: *Technical Bulletin*. CPCRI, Kasaragod, India, 34 pp.

Belavadi, V.V. and Parvathi, C. (1998) Estimation of honeybee colonies required for effective pollination in cardamom. In: *Proceedings of the National Symposium on Diversity of Social Insects and Other Arthropods and the Functioning Ecosystems, III Congress of IUSSI*, Mudigere, p. 30.

Belavadi, V.V., Venkateshalu, P. and Vivek, H.R. (1997) Significance of style in cardamom corolla tubes for honeybee pollination. *Current Science* 73, 287–290.

Bureau of Indian Standards (BIS) (2023) IS 1907: Specification for cardamom (capsules and seeds) (3rd rev.). Available at: https://www.bis.gov.in

Chakravarti, A.K. (1948) Multiplication of chromosome numbers in relation to speciation in Zingiberaceae. *Science and Culture* 14, 137–140.

Chandrasekhar, R. and Sampath Kumar, R. (1986) Karyological highlights on two cultivars of cardamom (*Elettaria cardamomum* Maton). *Journal of Cytology and Genetics* 21, 90–96.

Chaudhary, R. and Chandel, K.P.S. (1995) Studies on germination and cryo-preservation of cardamom (*Elettaria cardamomum* Maton) seeds. *Seed Science and Technology* 23, 235–240.

Chinnathambi, S., Peeran, M.F., Srinivasan, V., Sankar, S.M. and George, P. (2024) Optimizing mycorrhizal fungi application for improved nutrient uptake, growth, and disease resistance in cardamom seedlings (*Elettaria cardamomum* (L.) Maton). *Heliyon* 10(20), e39227. DOI: 10.1016/j.heliyon.2024.e39227.

Craig, P.S., Dujardin, B., Hart, A., Hernández-Jerez, A.F., *et al*, and European Food Safety Authority (EFSA) (2020) Cumulative dietary risk characterisation of pesticides that have acute effects on the nervous system. *EFSA Journal* 18(4), e06087. DOI: 10.2903/j.efsa.2020.6087.

Crop Production Guide: Horticulture Crops (2020) Published by Directorate of Horticulture and Plantation Crops, Chepauk, Chennai & Tamil Nadu Agricultural University Coimbatore-3 168–170.

Cyriac, A., Paul, R., Anupama, K., Senthil Kumar, R., Sheeja, T.E. *et al.* (2016) Isolation and characterization of genomic microsatellite markers for small cardamom (*Elettaria cardamomum* Maton) for utility in genetic diversity analysis. *Physiology and Molecular Biology of Plants* 22, 219–229.

Dandin, S.B., Madhusoodanan, K.J. and George, K.V. (1981) Cardamom descriptor. In: *Proceedings of the IV PLACROSYM*, CPCRI, Kasaragod, pp. 401–406.

Daneshi-Maskooni, M., Keshavarz, S.A., Qorbani, M., Mansouri, S., Alavian, S.M. *et al.* (2018) Green cardamom increases Sirtuin-1 and reduces inflammation in overweight or obese patients with non-alcoholic fatty liver disease: A double-blind randomized placebo-controlled clinical trial. *Nutrition & Metabolism* 15(1), 63. DOI: 10.1186/s12986-018-0297-4.

Daood, H.G. (ed.) (2008) Extraction of cardamom oil by supercritical carbon dioxide and sub-critical propane. *The Journal of Supercritical Fluids* 45(2), 208–213. DOI: 10.1016/j.supflu.2007.08.009.

Dattatray, S., Kamini, B. and Mohanlal, J. (2016) The Queen of Spices and Ayurveda: A brief review. *International Journal of Research in Ayurveda Pharm* 7(4), 1–6.

de Oliveira, E.R., de Souza, A.G. and dos Santos Rosa, D. (2025) From emulsions to solutions: Cardamom essential oil as a tool for packaging evolution. *Journal of Molecular Liquids* 429, 127549. DOI: 10.1016/j.molliq.2025.127549.

DGCI&S (2019) *Export-Import Data Bank*. Directorate General of Commercial Intelligence and Statistics, Ministry of Commerce, Government of India.

EFSA (European Food Safety Authority) (2019) Safety and efficacy of an essential oil from *Elettaria cardamomum* (L.) Maton when used as a sensory additive in feed for all animal species. *EFSA J* 17, e05721. DOI: 10.2903/j.efsa.2019.5721.

EFSA (European Food Safety Authority), Craig, P.S., Dujardin, B., Hart A., Hernández-Jerez, A.F., Hougaard Bennekou, S. *et al.* (2020) Cumulative dietary risk characterisation of pesticides that have acute effects on the nervous system. *EFSA J.* 18(4):e06087. DOI: 10.2903/j.efsa.2020.6087.

ESA (2018) Available at: https://www.esa-spices.org/media/attachments/2026/02/11/esa-quality-minima-document.pdf

El Sayed, S., Al-Thbity, S.M.K. and Gobouri, A.A. (2022) Impact of extraction methods on the chemical composition and biological activity of small cardamom essential oils. *Egyptian Journal of Chemistry* 65(12), 1–12. DOI: 10.21608/ejchem.2022.130386.5801.

Elguindy, N.M., Yacout, G.A., El Azab, E.F. and Maghraby, H.K. (2016) Chemoprotective effect of *Elettaria cardamomum* against chemically induced hepatocellular carcinoma in rats by inhibiting NF-κb, oxidative stress, and activity of ornithine decarboxylase. *South African Journal of Botany* 105, 251–258. DOI: 10.1016/j.sajb.2016.04.001.

European Commission (2025) *EU pesticides database*. Available at: https://ec.europa.eu/food/plant/pesticides/eu-pesticides-database/start/screen/mrls/details?lg_code=EN&pest_res_id_list=300&product_id_list=

FAO and WHO (2024) *Standard for Spices derived from Dried or Dehydrated fruits and berries – Small Cardamom*. Codex Alimentarius Standard, No. CXS 357-2024. Codex Alimentarius Commission, Rome.

FAOSTAT (2024) *Food and Agriculture Organization of the United Nations*. Available at: https://www.fao.org/faostat

Fukushima, S., Cohen, S.M., Eisenbrand, G., Gooderham, N.J., Guengerich, F.P. *et al.* (2020) FEMA GRAS assessment of natural flavor complexes: Lavender, Guaiac, Coriander-derived and related flavoring ingredients. *Food and Chemical Toxicology* 141, 111584. DOI: 10.1016/j.fct.2020.111584.

Gaikwad, A.B., Kumari, R., Yadav, S., Rangan, P., Wankhede, D.P. *et al.* (2023) Small cardamom genome: Development and utilization of microsatellite markers from a draft genome sequence of *Elettaria cardamomum* Maton. *Frontiers in Plant Science* 14, 1161499. DOI: 10.3389/fpls.1161499.

Gangadharan, S., Augustin, L.E. and Thanga Pushpam, R. (2025) Growth performance and instability in Indian cardamom exports: A time series analysis. *Journal of Experimental Agriculture International* 47(9), 114–120. DOI: 10.9734/jeai/2025/v47i93732.

George, M. and Cherian, E. (2017) Emergent global marketing challenges for Kerala cardamom producers vis-à-vis the role of the spices board of India. *International Journal of Community Development & Management Studies* 1, 39–62.

Gopal, R., Chandraswamy, D.W. and Nayar, N.K. (1990) Correlation and path analysis in cardamom. *Indian Journal of Agricultural Sciences* 60, 240–242.

Gopal, R., Chandraswamy, D.W. and Nayar, N.K. (1992) Genetic basis of yield and yield components in cardamom. *Journal of Plantation Crops* 20, 230–232.

Gregory, P.J. (1936) The floral morphology and cytology of *Elettaria cardamomum* Maton. *Journal of the Linnean Society* 50, 363–391.

Grover, M. (2021) Ayurvedic and therapeutic significance of queen of spices, *Elettaria cardamomum* Maton (choti elaichi). *World J Pharmaceutical Med Res* 7(8), 290–300.

Gurumurthy, B.R., Chandrappa, H.M., Parvathi, C., Raju, B. and Veerabadraiah, S. (1996) Screening of cardamom clones for drought tolerance. In: Satheesan, K.V. (ed.) *Proceedings of the National Seminar on Drought Management in Plantation Crops*, Kottayam, Kerala, pp. 25–29.

Hamdan, S., Daood, H.G., Toth-Markus, M. and Illés, V. (2008) Extraction of cardamom oil by supercritical carbon dioxide and sub-critical propane. *The Journal of Supercritical Fluids* 44(1), 38–44. DOI: 10.1016/j.supflu.2007.08.009.

Hegde, R. and Korikanthimath, V.S. (1996) Agronomic approaches for drought management in cardamom. In: Satheesan, K.V. (ed.) *Proceedings of the National Seminar on Drought Management in Plantation Crops*, Kottayam, Kerala, pp. 75–79.

Herrera-González, M.P., Zamora-Jerez, A., Cifuentes-Velasquez, R., Arévalo-Rodríguez, L.A. and Pereira-Lorenzo, S. (2024) Comprehensive evaluation and selection of cardamom (*Elettaria cardamomum* (L.) Maton) germplasm using morphological traits. *Plants* 13(19), 2786. DOI: 10.3390/plants13192786.

Hoberman, A.M. (1989) Reproductive and Developmental Toxicity Screening Test of B9 (Cardamom Oil) Administered Orally via Gavage to Crl:CD (SD) BR Female Rats. *Unpublished Study Report*, Study no.412-005. Horsham, PA: Argus Research Laboratories, Inc.

Honnappa, A., Shivakumar, M.S., Akshitha, H.J., Mohammed, F.P., Balaji, R.M. *et al.* (2025) Morphological variability and disease resistance levels in cardamom (*Elettaria cardamomum* Maton) field gene bank accessions. *Annual Research & Review in Biology* 40(3), 73–81.

Hrideek, T.K., Radhakrishnan, V.V., Mohanan, K.V., Kuruvilla, K.M., Madhusoodanan, K.J. *et al.* (2011) Genetic divergence in some elite landraces of small cardamom (*Elettaria cardamomum* Maton). *Journal of Plantation Crops* 39, 201–202. Available at: http://ijcdms.org/Volume01/v1p039-061George3465.pdf

IISR (2022) *Cardamom – extension pamphlet.* ICAR-Indian Institute of Spices Research, Kozhikode, India, p. 28.

IISR Annual Report (2021) *Integrated nematode management in spices.* ICAR–Indian Institute of Spices Research, Kozhikode.

Indhushree, A. and Kuruvila, A. (2019) Performance of small cardamom export from India. *Journal of Tropical Agriculture* 57(2), 122–131.

IPGRI (1994) *Descriptors for cardamom (Elettaria cardamomum Maton).* International Plant Genetic Resources Institute (IPGRI), Rome, pp. 27–41.

ITC Trade Map (2023) *International Trade Centre, Geneva.* Available at: https://www.trademap.org

Jacob, T.K., Senthil Kumar, C.M., Devasahayam, S., D'silva, S., Senthil Kumar, R. *et al.* (2020) Plant morphological traits associated with field resistance to cardamom thrips (*Sciothrips cardamomi*) in cardamom (*Elettaria cardamomum*). *Annals of Applied Biology*177, 143–151. DOI: 10.1111/aab.12592.

Jadav, K.D. and Mehta, B.M. (2018) Cardamom: Chemistry, medicinal properties, applications in dairy and food industry. *Journal of Dairy Science and Technology* 7(3), 9–19.

Jamil, B., Habib, H., Abbasi, S.A. and Imran, M. (2016) Encapsulation of cardamom essential oil in chitosan nano-composites: *In-vitro* efficacy on antibiotic-resistant bacterial pathogens and cytotoxicity studies. *Frontiers in Microbiology* 7, 1580. DOI: 10.3389/fmicb.2016.01580.

Jegadeesan, M. and Miyazaki, H. (2025) Cardamom cultivation in Idukki, India: Practices, challenges, and historical insights. *International Journal of Horticulture and Food Science* 7(1), 82–88. DOI: 10.33545/26631067.2025.v7.i1b.246.

Katekar, V.P., Rao, A.B. and Sardeshpande, V.R. (2023) A cutting-edge assessment of recent advancements in essential oils extraction technologies for energy and ecological sustainability: A comprehensive review. *Journal of Food Process Engineering* 46(9), e14414. DOI: 10.1111/jfpe.14414.

KAU (2017) The package of practices: Crops. Directorate of Research Vellanikkara, KAU Post, Thrissur, Kerala, India, 360 pp.

Kazemi, S., Yaghooblou, F., Siassi, F., Rahimi Foroushani, A., Ghavipour, M. *et al.* (2017) Cardamom supplementation improves inflammatory and oxidative stress biomarkers in hyperlipidemic, overweight, and obese pre-diabetic women: A randomized double-blind clinical trial. *Journal of the Science of Food and Agriculture* 97(15), 5296–5301. DOI: 10.1002/jsfa.8414.

Khan, Z.A., Kumar, R. and Dasgupta, I. (2022) CRISPR/Cas-Mediated resistance against viruses in plants. *Interntional Journal of Molecular Sciences* 23(4), 2303. DOI: 10.3390/ijms23042303.

Khattab, A.A., Tawfek, A.M., Abo-EL-Sooud, K., Ahmed, K.A., El-Gendy, A.E.N. *et al.* (2020) *Elettaria cardamomum* essential oil rescues paracetamol-induced hepatorenal damage via modulating oxidative stress in rats. *Advances in Animal and Veterinary Sciences* 8(S2), 24–33.

Khulbe, D. (2019) Integrating soil solarization and seed biopriming to manage seedling damping-off in flower nurseries. *International Journal of Current Microbiology and Applied Sciences* 8(2), 1456–1469. DOI: 10.20546/ijcmas.2019.802.169.

Kiralan, M., Ketenoglu, O. and Kiralan, S.S. (2023) Composition and functional properties of cardamom essential oil. In: Ramadan, M.F. (ed.) *Cardamom (Elettaria Cardamomum): Production, Processing and Properties*. Springer, Cham, pp. 105–118. DOI: 10.1007/978-3-031-35426-7_8.

Korikanthimath, V.S. (1993) Harvesting and on-farm processing of cardamom. In: *Proceedings of the National Seminar on Post-Harvest Technology of Spices*, Indian Society for Spices (ISS), Calicut, pp. 62–68.

Korikanthimath, V.S. and Mulge, R. (2005) Studies on panicle and capsule characters in elite clones of cardamom. *Indian Journal of Horticulture* 62(4), 367–369.

Korikanthimath, V.S. and Venugopal, M.N. (1989) High production technology in cardamom. Tech Bull, National Research Centre for Spices, Calicut, Kerala.

Korikanthimath, V.S., Mulge, R., Hegde, R. and Hosmani, M.M. (1999) Correlation between yield and yield parameters in cardamom (*Elettaria cardamomum*). *Journal of Medicinal and Aromatic Plants* 21, 700–701.

Korikanthimath, V.S., Prasath, D. and Rao, G. (2000a) Medicinal properties of cardamom (*Elettaria cardamomum*). *Journal of Medicinal and Aromatic Plant Sciences* 22, 683–685.

Korikanthimath, V.S., Prasath, D. and Mohammed Sayed, A.A. (2000b) Evaluation of cardamom accessions in relation to yield, recovery and oil content. *Indian Journal of Horticulture* 57(3), 273–275.

Korikanthimath, V.S., Ankegowda, S.J. and Akshitha, H.J. (2024) Cardamom [ISBN 978-981-19-3727-9; ISBN 978-981-19-3728-6 (eBook)]. In: Ravindran ., P.N., Sivaraman, K., Devasahayam, S. and Nirmal Babu, K. (eds) *Handbook of Spices in India: 75 Years of Research and Development*. Springer Nature Singapore Pte Ltd, Singapore. DOI: 10.1007/978-981-19-3728-6.

Krishnamurthy, K., Khan, M.M. and Avadhani, K.K. (1989) *Three decades of cardamom research at R.R.S., Mudigere*. Technical Bulletin, University of Agricultural Sciences, Bangalore, India.

Kumar, K.B., Prakashkumar, P., Balachandran, S.M. and Iyer, R.D. (1985) Development of clonal plantlets from immature panicles of cardamom. *Journal of Plantation Crops* 13, 31–34.

Kumari, A., Singla, R., Guliani, A. and Yadav, S.K. (2014) Nanoencapsulation for drug delivery. *EXCLI Journal* 13, 265.

Kuriakose, G., Sinu, P.A. and Shivanna, K.R. (2009) Domestication of cardamom (*Elettaria cardamomum*) in Western Ghats, India: Divergence in productive traits and a shift in major pollinators. *Annals of Botany* 103, 727–733. DOI: 10.1093/aob/mcn262.

Leela, N.K., Prasath, D. and Venugopal, M.N. (2008) Essential oil composition of selected cardamom genotypes at different maturity levels. *Indian Journal of Horticulture* 65(3), 366–369.

Madhusoodanan, K.J., Pradip Kumar, K. and Ravindran, P.N. (2002) Botany, crop improvement and biotechnology of cardamom. In: Ravindran, P.N. and Madhusoodanan, K.J. (eds) *Cardamom: The Genus Elettaria*. Taylor & Francis, Cardamom, pp. 11–68.

Mahindru, S.N. (1982) *Spices in Indian life*. Sultanchand & Sons, New Delhi.

Malhotra, E.V., Kamalapriya, M., Bansal, S., Meena, D.P.S. and Agrawal, A. (2021) Improved protocol for micropropagation of genetically uniform plants of commercially important cardamom (*Elettaria cardamomum* Maton). *In Vitro Cellular & Developmental Biology-Plant* 57, 409–417. DOI: 10.1007/s11627-020-10131-7.

Manohari, C., Backiarani, S., Jebasingh, T., Somanath, A. and Usha, R. (2008) Efficient plant regeneration in small cardamom (*Elettaria cardamomum* Maton), through somatic embryogenesis. *Indian Journal of Biotechnology* 7, 407–409.

Mayne, W.W. (1942) *Report on cardamom cultivation in South India*, Miscellaneous Bulletin no.50. Imperial Council of Agricultural Research, New Delhi.

Mehyar, G., Al-Ismail, K., Al-Ghizzawi, H. and Holley, R. (2014) Stability of cardamom (*Elettaria cardamomum*) essential oil in microcapsules made of whey protein isolate, guar gum, and carrageenan. *Journal of Food Science* 79. DOI: 10.1111/1750-3841.12652.

Milian, L.S. (2014) Cardamom - The 3Gs -- Green Gold of Guatemala. USDA FAS GAIN Report GT-1404. USDA Foreign Agricultural Service. Available at: fas.usda.gov (accessed 22 November 2025).

Ming, L., Xiao, Z., Nie, Z., Xiao, N., Wu, H. *et al.* (2025) Encapsulation of essential oil in cyclodextrin: An exploration of rational design, molecular interactions, and activity assessment. *Industrial Crops and Products* 226, 120611. DOI: 10.1016/j.indcrop.2025.120611.

Ministry of Commerce (2025) *Trade statistics*. Department of Commerce, Ministry of Commerce and Industry, Government of India. Available at: https://tradestat.commerce.gov.in/

Ministry of Commerce and Industry (2025) *Market development assistance (MDA) scheme*. Available at: https://www.commerce.gov.in/international-trade/trade-promotion-programmes-and-schemes/trade-promotion-programme-focus-cis/market-development-assistance-mda-scheme

Misra, S. (ed.) (2011) *Kaviraj Govind Dasasen: Bhaiṣhajya Ratnavali, with 'Siddhiprada' Hindi commentrary*. Chaukha Surabharati Prakashan, Varanasi.

Moench, M. (1991) Politics of deforestation: Case study of the Cardamom Hills of Kerala. *Economic and Political Weekly* 26(4), PE47–PE60.

Mohemmad, R.N., Ahmed, B.S., Ahmed, V.A., Ismail, S.M. and Rahman, H.S. (2024) Anti-inflammatory, antioxidant, and gastro-protective actions of cardamom essential oil-loaded nanostructured lipid carrier in ethanol-induced gastric ulcer. *Natural Product Communications* 19(8), 1–10. DOI: 10.1177/1934578X241272492.

Morsy, S.M. (2015) Composition and functional properties of cardamom essential oil. In: Rahman, M.M.A.A. and El-Morsy, M.M.A.F. (eds) *Essential Oils: Extraction and Applications*. Springer, pp. 145–168.

Murugan, M., Dhanya, M.K. and Deepthy, K.B. (2016) *Compendium on Cardamom*, 2nd edn. Kerala Agricultural University, Cardamom Research Station, Pampadumpara, Kerala, 66 pp.

Murugan, M., Ravi, R., Anandhi, A., Kurien, S. and Dhanya, M.K. (2017) Pesticide use in Indian cardamom needs change in cultivation practices. *Current Science* 113(6), 1058–1063.

Murugan, M., Dhanya, M.K., Ashokkumar, K., Sathyan, T. and Surya, R. (2019) Changes of enzyme activities and phytochemical constituents in small cardamom capsules caused by the infestation of thrips, *Sciothrips cardamomi* (Ramk). *Research Journal of Biotechnology* 14(10), 113–116.

Murugan, M., Ashokkumar, K., Dhanya, M.K. and Subbiah, A. (2021) History of land and agriculture in the Indian Cardamom Hills. In: Ashokkumar, K., Murugan, M. and Dhanya, M.K. (eds) *Cardamom (Elettaria Cardamomum Maton)*. Nova Science Publishers, New York, pp. 1–20.

Murugan, K., Ramesh, P. and Devi, R. (2024) Analyzing the performance of small cardamom in India. *Economic Affairs* 69(1).

Muthusamy, R.B., Sivaranjani, R., Senthil Kumar, C.M., Punithavalli, M., Asangi, H. *et al.* (2025) Impact of thrips (*Sciothrips cardamomi*) damage on volatile composition and other attributes of cardamom capsules (*Elettaria cardamomum*). *Annals of Applied Biology* 1–14. DOI: 10.1111/aab.70077.

Nadgauda, R.S., Mascareenhas, A.F. and Madhusoodanan, K.J. (1983) Clonal multiplication of cardamom (*Elettaria cardamomum* Maton) by tissue culture. *Journal of Plantation Crops* 11(1), 60–66.

Nadiya, F., Anjali, N., Jinu, T., Gangaprasad, A. and Sabu, K.K. (2017) Transcriptome profiling of *Elettaria cardamomum* (L.) Maton (small cardamom). *Genomics Data* 11, 102–103. DOI: 10.1016/j.gdata.2016.12.013.

Nadiya, F., Anjali, N., Jinu, T., Gangaprasad, A. and Sabu, K.K. (2018) Genome-wide differential expression profiling in wild and cultivar genotypes of cardamom reveals regulation of key pathways in plant growth and development. *Agri Gene* 8, 18–27. DOI: 10.1016/j.aggene.2018.03.002.

Nadukeri, S., Umesha, K., Hanumanthappa, M., Kolakar, S. and Venukumar, S. (2018) Constraints in cultivation of cardamom in Karnataka: Review. *Indian Journal of Arecanut, Spices and Medicinal Plants* 20(2), 52–58.

Nadukeri, S., Umesha, K., Naik, M.K., Srinivasa, V., Ankegowda, S.J. *et al.* (2020) Evaluation of working collections of cardamom (*Elettaria cardamomum* Maton) for morphological attributes under hill zone of Karnataka. *International Journal of Chemical Studies* 8(6), 2215–2219.

Nair, R.R., Thomas, S. and Sabu, T. (2019) Integrated management of pests and diseases in small cardamom (*Elettaria cardamomum* Maton). *Indian Journal of Spices* 26(2), 85–95.

Nirmal Babu, K., Geetha, S.P., Minoo, D., Ravindran, P.N. and Peter, K.V. (1999a) *In vitro* conservation of cardamom (*Elettaria cardamomum* Maton) germplasm. *Plant Genetic Resources Newsletter (IPGRI)* 119, 41–45.

Nirmal Babu, K., Geetha, S.P., Minoo, D., Ravindran, P.N. and Peter, K.V. (1999b) *In vitro* conservation of germplasm. In: Ghosh, S.P. (ed.) *Biotechnology and Its Application in Horticulture*. Narosa Publishing House, New Delhi, pp. 106–129.

Nishteswar, N. and Vidyanath, R. (eds) (2023) *Sahasrayogam,* Banaras Ayurveda Series 18. Chowkhamba Sanskrit Series Office, Varanasi.

Nnadi, N.E. and Carter, D.A. (2021) Climate change and the emergence of fungal pathogens. *PLoS Pathogens* 17(4), e1009503. DOI: 10.1371/journal.ppat.1009503.

Noktehsanj, M., Molaveisi, M., Mohammadi, M. and Noghabi, M. (2020) Chemical composition, antioxidant potential, and antimicrobial activity of *Elettaria cardamomum* essential oil. *Infection Epidemiology and Microbiology* 6(1), 51–62.

Noumi, E., Snoussi, M., Alreshidi, M.M., Rekha, P.D., Saptami, K. *et al.* (2018) Chemical and biological evaluation of essential oils from cardamom species. *Molecules* 23(11), 2818. DOI: 10.3390/molecules23112818.

Padmini, K., Venugopal, M.N. and Korikanthimath, V.S. (1999) Biodiversity and conservation of cardamom (*Elettaria cardamomum* Maton). In: *Biodiversity, Conservation and Utilization of Spices, Medicinal and Aromatic Plants*. Indian Institute of Spices Research (IISR), Calicut, pp. 73–78.

Padmini, K., Venugopal, M.N. and Ankegowda, S.J. (2000a) Heterosis for seedling characters in cardamom (*Elettaria cardamomum* Maton). *Journal of Spices and Aromatic Crops* 9(2), 149–153.

Padmini, K., Venugopal, M.N. and Sasikurnar, B. (2000b) Performance of hybrids, open pollinated progenies and inbreds of cardamom (*Elettaria cardamomum* Maton) under nursery conditions. *Indian Journal of Agricultural Sciences* 70(8), 550–551.

Pai, R.M. (1965) The floral anatomy of *Elettaria cardamomum* Maton. *New Physiology* 64, 187–194.

Panchaksharappa, M.G. (1966) Embryological studies in some members of Zingiberaceae II: *Elettaria cardamomum, Hitchenia caulina* and *Zingiber macrostachyum*. *Phytomorphology* 16, 412–417.

Parameswar, N.S. (1973) Floral biology of cardamom (*Elettaria cardamomum* Maton). *Mysore Journal of Agricultural Sciences* 7, 205–213.

Parameswar, N.S. and Venugopal, R. (1974) Study of flowering and anthesis in cardamom (*Elettaria cardamomum* Maton). *Mysore Journal of Agricultural Sciences* 6, 356–361.

Pattanashetti, H.V. and Prasad, A.B.N. (1972) Blossom biology, pollination and fruit set in cardamom (*Elettaria cardamomum* Maton). In: *Third International Symposium on Subtropical and Tropic Crops*. pp. 262–268.

Paul, J., Yadav, B.S.P., Kumar, P.K. and Rema Shree, A.B. (2024) A comparative study of essential oil components of selected promising small cardamom (*Elettaria cardamomum* (L.) Maton) landraces in Cardamom Hill Reserve (CHR) of South India. *Journal of Applied and Natural Science* 16(1), 123–132. DOI: 10.31018/jans.v16i1.5302.

Poulson, A.D., Fer, T., Marasinghe, L.D.K., Sabu, M., Hughes, M. *et al.* (2024) The cardamom conundrum resolved: Recircumscription and placement of *Elettaria* in the only pantropically distributed ginger lineage. *Taxon* 73(5), 1187–1213. DOI: 10.1002/tax.13242.

PPV&FRA (2007) *Guidelines for the conduct of test for distinctiveness, uniformity and stability on small cardamom (Elettaria cardamomum Maton)*. Protection of Plant Varieties and Farmers' Rights Authority, Government of India.

Prabhakaran Nair, K.P. (2006) The agronomy and economy of cardamom (*Elettaria cardamomum* M.): The "Queen of Spices". *Advances in Agronomy* 91, 179–471. DOI: 10.1016/S0065-2113(06)91004-9.

Prasath, D. and Venugopal, M.N. (2004) Genetic diversity and conservation of cardamom (*Elettaria cardamomum* Maton) in India. *Plant Genetic Resources Newsletter* 138, 55–60.

Prasath, D. and Venugopal, M.N. (2009) Compound inflorescence cardamom (*Elettaria cardamomum* (L.) Maton) in India. *Genetic Resources and Crop Evolution* 56, 749–753. DOI: 10.1007/s10722-009-9440-4.

Prasath, D., Venugopal, M.N. and Korikanthimath, V.S. (2001) Variability in cardamom (*Elettaria cardamomum* Maton). *Indian Journal of Plant Genetic Resources* 14(2), 217–218.

Prasath, D., Venugopal, M.N. and Senthil Kumar, R. (2009a) Evaluation of high yielding cardamom selection in Karnataka for yield and quality. *Indian Journal of Agricultural Sciences* 79(3), 207–209.

Prasath, D., Venugopal, M.N., Senthilkumar, R. and Leela, N.K. (2009b) Hybrid performance for yield and yield components in cardamom (*Elettaria cardamomum* Maton). *Euphytica* 168(1), 49–60. DOI: 10.1007/s10681-008-9878-z.

Prasath, D., Venugopal, M.N. and Parthasarathy, V.A. (2010) Inheritance of Cardamom Mosaic Virus (CdMV) resistance in cardamom. *Scientia Horticulturae* 125, 539–541. DOI: 10.1016/j.scienta.2010.03.018.

Praveena, R., Biju, C.N., Senthilkumar, R., Darshana, C.N. and Jashmi, K.C. (2013) Preliminary evaluation of cardamom accessions against leaf blight (Chenthal disease). *Indian Phytopathology* 66(1), 112–113.

Preethy, T.T., Dhanya, M.K., Aswathy, T.S., Sathyan, T., Backiyarani, S. *et al.* (2022) Assessment of genetic diversity of small cardamom (*Elettaria cardamomum* M) in India. *Journal of Plant Genetic Resources* 35(2), 169–177.

Priyadarshan, P.M., Kuruvilla, K.M., Madhusoodanan, K.J. and Naidu, R. (1992) Effects of vivid media protocols and genetic specificity of micropropagation of cardamom (*Elettaria cardamomum* Maton). In: Subba Rao, N.S., Balachandram, C. and Ramakrishnan, S.V. (eds) *New Trends in Biotechnology*. Oxford and IBH Publishing, pp. 109–118.

Pujiarti, R. and Kusumadewi, A. (2020) Chemical compounds, physicochemical properties, and antioxidant activity of *A. cardamomum* leaves and rhizomes oils on different distillation time. *Wood Research Journal* 11(1), 35–40.

Qiblawi, S., Kausar, M.A., Shahid, S.M.A., Saeed, M. and Alazzeh, A.Y. (2020) Therapeutic interventions of cardamom in cancer and other human diseases. *Journal of Pharmaceutical Research International* 32(22), 74–84. DOI: 10.9734/JPRI/2020/v32i2230774.

Rahman, M.M., Alam, M.N., Ulla, A., Sumi, F.A., Subhan, N. *et al.* (2017) Cardamom powder supplementation prevents obesity, improves glucose intolerance, inflammation and oxidative stress in liver of high carbohydrate high fat diet induced obese rats. *Lipids in Health and Disease* 16(1), 151. DOI: 10.1186/s12944-017-0539-x.

Raissa, R., Amalia, W.C., Ayurini, M., Khumaini, K. and Ratri, P.J. (2020) The optimization of essential oil extraction from java cardamom. *Journal of Tropical Pharmacy and Chemistry* 5(2), 125–129. DOI: 10.25026/jtpc.v5i2.257.

Raj, N.M., Murugan, M. and Joseph, R.C. (2000) Evaluation of cardamom germplasm. *Journal of Spices and Aromatic Crops* 9(1), 55–56.

Rajkumar, A.J., Backiyarani, S. and Sivakumar, G. (2007) *Gleanings in cardamom*. Kerala Agricultural University, Cardamom research station, Pampadumpara, p. p–81.

Rao, N.S., Narayanaswamy, S., Chacko, E.K. and Swamy, R.D. (1982) Regeneration of plantlets from callus of *Elettaria cardamomum* Maton. *Proceedings of Indian Academy of Science (Plant Science)* 91, 37–41.

Ravindran, P.N. and Madhusoodanan, K.J. (eds) (2002) *Cardamom: The Genus Elettaria*, 1st edn. CRC Press. DOI: 10.1201/9780203216637.

Reghunath, B.R. and Bajaj, Y.P.S. (1992) Micropropagation of cardamom (*Elettaria cardamomum* Maton). In: Bajaj, Y.P.S. (ed.) *High-Tech and Micropropagation III*. Springer, Berlin, Heidelberg, pp. 175–198.

Remya, J.S., Dhanya, M.K., Murugan, M., Rafi, N. and Sreekumar, A. (2025) Screening Indian cardamom accessions for resistance to major pathogens and correlating responses with biochemical defense parameters. *Journal of Advances in Biology & Biotechnology* 28(8), 1067–1083. DOI: 10.9734/jabb/2025/v28i82780.

Grand View Research. (2025) *Green cardamom market report: Market size & forecast.* Available at: https://www.grandviewresearch.com/industry-analysis/green-cardamom-market-report#:~:text=Market%20Size%20&%20Forecast,and%20contributing%20to%20market%20expansion

Sabu, T. and Menon, S. (2020a) Eco-friendly management of capsule rot in cardamom using bioagents. *Spice Research Frontier* 11(1), 33–39.

Sabu, T. and Menon, S. (2020b) Pheromone-based integrated management of shoot borer in cardamom plantations. *Journal of Plantation Crops* 48(3), 217–223.
Sannagoudar, M.S., Patil, R.H., Rajanna, G.A., Ghosh, A., Singh, A.K. *et al.* (2023) Rising temperature coupled with reduced rainfall will adversely affect yield of kharif sorghum genotypes. *Current Science* 124(8), 1–7. DOI: 10.18520/cs/v124/i8.
Sarathkumaran, S.J., Paciasothy, E.V. and Jansz, E.R. (1985) Some studies on the effect of maturity and storage on the chlorophyll content and essential oils of the cardamom fruit (*Elettaria cardamomum* Maton). *Journal of the Science of Food and Agriculture* 36(6), 491–498.
Savan, E.K. (2023) Encapsulation of Cardamom Extracts. In: *Cardamom (Elettaria Cardamomum): Production, Processing and Properties*. Springer International Publishing, Cham, pp. 255–276. DOI: 10.1007/978-3-031-35426-7_19.
Senthil Kumar, R., Prasath, D., Aravind, S., Leela, N.K., Venugopal, M.N. *et al.* (2017) Evaluation of elite cardamom (*Elettaria cardamomum*) genotypes for yield, quality, and resistance to cardamom mosaic virus. *Journal of Plant Crops* 45(2), 129–134.
Senthil Kumar, R., Sharon, A., Biju, C.N. and Ankegowda, S.J. (2018) Characterization of genotypes of small cardamom (*Elettaria cardamomum* Maton) for yield parameters and disease resistance. *Journal of Plant Crops* 46(2), 143–146.
Shahidullah, M., Janarthan, M. and Khan, M.S. (2017) Evaluation of cardioprotective activity of maceration extract of *Elettaria cardamomum* in doxorubicin-induced cardiotoxicity in rats. *Indian Journal of Research in Pharmacy and Biotechnology* 5(6), 366–370.
Sharma, P. (ed.) (1978) Sodhala. In: *ŚodhālaNighaṇṭu*. Oriental Institute, Baroda.
Sharma, P. (ed.) (2005) *Dhanvantari Nighaṇṭu*, 4th Edition. Chaukhambha Orientalia, Varanasi.
Sharma, A.K. and Bhattacharyya, N.K. (1959) Cytology of several members of Zingiberaceae and the study of the inconsistency of their chromosome complements. *La Cellule* 59, 297–346.
Sharon, A., Biju, C.N., Ankegowda, S.J., Shivakumar, M.S., Senthil Kumar, R. *et al.* (2020) Screening small cardamom (*Elettaria cardamomum* Maton) field gene bank accessions for phenotypic characters, yield potential and disease resistance. *Journal of Spices and Aromatic Crops* 29(2), 105–112.
Shasthri, L. (ed.) (2010) *Yogaratnākara with commentary "Vidhyotani"*. Chaukhambha Prakashana, Varanasi.
Shashank, R., Doddabasappa, B., Vasudev, K., Chakravarthy, A.K. and Hiroshi, H. (2015). Molecular Characterization and Management of Shoot and Fruit Borer Conogethes punctiferalis Guenee (Crambidae: Lepidoptera) Populations Infesting Cardamom, Castor and Other Hosts. In: Chakravarthy, A. (eds) *New Horizons in Insect Science: Towards Sustainable Pest Management*. Springer, New Delhi. DOI: 10.1007/978-81-322-2089-3_20.
Singletary, K. (2022) Cardamom: Potential health benefits. *Nutrition Today* 57(1), 38–49.
Sinu, P.A. and Shivanna, K.R. (2007) Pollination ecology of cardamom (*Elettaria cardamomum*) in the Western Ghats, India. *Journal of Tropical Ecology* 23, 493–496.
Sivanandan, P., Narayanan, D. and Narayanan, N.K. (1986) Land hunger and deforestation: A case study of Cardamom Hills of Kerala. *Economic and Political Weekly* 21(13), 546–550.
Sivaraman, K., Thankamani, C.K. and Srinivasan, V. (2023) Crop diversification: Cropping/system approach for enhancing farmers' income. In: *Handbook of Spices in India: 75 Years of Research and Development*. pp. 3847–3926. DOI: 10.1007/978-981-19-3728-6_61.

Sontakke, M.D., Syed, H.M. and Sawate, A.R. (2018) Studies on extraction of essential oils from spices (Cardamom and Cinnamon). *International Journal of Chemical Studies* 6(2), 2787–2789.

Souza, A.G., Ferreira, R.R., de Oliveira, E.R., Kato, M.M., Mitra, S.K. *et al.* (2022) Chemical stabilization behind cardamom pickering emulsion using nanocellulose. *Polysaccharides* 3(1), 200–216. DOI: 10.3390/polysaccharides3010010.

Spices Board. (2025a) *Collaborative projects on traceability, sustainability, biodiversity and certification to enhance export of spices.*

Spices Board. (2025b) *Spices production data.* Spices Board, Ministry of Commerce & Industry, Government of India. Available at: https://www.indianspices.com/sites/default/files/Major%20spice%20state%20wise%20area%20production%202024-25%20web.pdf

Spices Board. (2025c) *Major item/country-wise export of spices from India.* Spices Board, Ministry of Commerce & Industry, Government of India. Available at: https://www.indianspices.com/sites/default/files/majoritemcountrywise%20export%202025%20web.pdf

Spices Board (2025d) *Market development assistance (MDA) scheme.* Spices Board, Ministry of Commerce & Industry, Government of India. Available at: https://indianspices.com/sites/default/files/mda.pdf

Spices Board (2025e) *SPICED scheme.* Spices Board, Ministry of Commerce & Industry, Government of India. Available at: https://www.indianspices.com/box5_programmes_schemes.html

Spices Board. (2025f) *List of Geographical Indication (GI) tags registered for spices.* Spices Board, Ministry of Commerce & Industry, Government of India. Available at: https://www.indianspices.com/sites/default/files/List_of_GItags_for_spices_BJ.pdf

Srinivasan, V., Alagupalamuthirsolai, M., Singh, R. and Dinesh, R. (2024) Precision agriculture technologies in spices. In: Ravindran, P.N., Sivaraman, K., Devasahayam, S. and Babu, K.N. (eds) *Handbook of Spices in India: 75 Years of Research and Development.* Springer, Singapore. DOI: 10.1007/978-981-19-3728-6_62.

Subbiah, A., Murugan, M., Rameshkumar, A., Jagadeesan, R. and Prabhu, M. (2008) A case study: Nutrient management in small cardamom (*Elettaria cardamomum*). *The Asian Journal of Horticulture* 3(2), 449–455.

Sudharshan, M.R. (1987) Colchicine-induced tetraploids in cardamom (*Elettaria cardamomum* Maton). *Current Science* 56(1), 36–37.

Sudharshan, M.R. (1989) Induced polyploids in cardamom. *Journal of Plant Crops* 16, 365–369.

Sudharsan, M.R., Kuruvilla, K.M. and Madhusoodanan, K.J. (1988) A key to identification of types of cardamom. *PLACROSYM* VIII, 52–55.

Tan, C.X., Tan, S.S. and Tan, S.T. (2023) Composition and functional properties of cardamom seeds. In: Ramadan, M.F. (ed.) *Cardamom: Production, Processing and Properties.* Springer, pp. 93–104. DOI: 10.1007/978-3-031-35426-7_7.

Thankamani, C.K., Srinivasan, V., Remya, J.S., Murugan, M., Dhanya, M.K. *et al.* (2024) Organic farming of spices: Concepts, issues, and strategies. In: *Handbook of Spices in India: 75 Years of Research and Development.* pp. 3949–4054. DOI: 10.1007/978-981-19-3728-6_63.

Thomas, J., Suseela Bhai, R.., Dhanapal, K. and Vijayan, A.K. (1997) *Integrated management of rot diseases of cardamom.* In: *Paper Presented in Integrated Plant Disease Management for Sustainable Agriculture,* New Delhi, November 10–15, .

Thomas, L., Rajeev, P. and Sanil, P.C. (2019) Trade competitiveness and export performance of Indian cardamom. *Journal of Spices and Aromatic Crops* 28(1), 34–42. DOI: 10.25081/josac.2019.v28.i1.5742.

Thorp, K.R., DeJonge, K.C., Kaleita, A.L., Batchelor, W.D. and Paz, J.O. (2008) Methodology for the use of DSSAT models for precision agriculture decision support. *Computers and Electronics in Agriculture* 64(2), 276–285. DOI: 10.1016/j.compag.2008.05.022.

Tridge (2025) Cardamom from Guatemala. Tridge Market Guides. Available at: https://www.tridge.com/tridge-woods/cardamom/GT

Tripathi, I. (ed.) (2006) *Narhari Pandita: Rājanighaṇṭu*, 4th ed. Chowkhambha Krishnadas Academy, Varanasi.

Tripathi, B. (ed.) (2009) Vagbhata. In: *Ashtanga Hridayam (Nirmala Hindi Commentary)*, Reprint. Chaukhamba Sanskrit pratishthan, Delhi.

Tyagi, R.K., Goswami, R., Sanayaima, R., Singh, R., Tandon, R. *et al.* (2009) Micropropagation and slow growth conservation of cardamom (*Elettaria cardamomum* Maton). In: *In Vitro Cellular & Development Biology-Plant*, Vol. 45. pp. 721–729.

Vaidya, S. and Rathod, N. (2014) Phytochemical constituents and pharmacological activities of cardamom (*Elettaria cardamomum*): A review. *International Journal of Pharmacy and Pharmaceutical Sciences* 6(6), 52–55.

Vaidya, E., Kamalapriya, M., Bansal, S., Meena, D.P.S. and Agrawal, A. (2020) Improved protocol for micropropagation of genetically uniform plants of commercially important cardamom (*Elettaria cardamomum* Maton). *In Vitro Cellular & Developmental Biology - Plant* 57(4), 409–417.

Varrier, K.R. (2005) *Arogyakalpadruma*. Sulabha books Thrissur, Kerala.

Vellaram, S. (2024) Spices Board imposes restrictions on cardamom re-pooling. *The Hindu*. Available at: https://www.thehindu.com/news/national/kerala/spices-board-imposes-restrictions-on-cardamom-re-pooling/article68259181.ece

Venugopal, M.N. (1999) Natural disease escapes as source of resistance against cardamom mosaic virus causing katte disease of cardamom (*Elettaria cardamomum* Maton). *Journal of Spices and Aromatic Crops* 8, 145–151.

Venugopal, M.N. and Padmini, K. (1999) Studies on crossability between elite types of cardamom (*Elettaria cardamomum* Maton). *Journal of Spices and Aromatic Crops* 8(2), 185–187.

Venugopal, M.N., Prasath, D. and Mulge, R. (2006) IISR Avinash-a rhizome rot resistant and high yielding variety of cardamom (*Elettaria cardamomum* Maton). *Journal of Spices and Aromatic Crops* 15(1), 14–18.

World Trade Organization (2022) *The SPS Agreement, the SPS Committee, and pesticide MRLs*. Available at: https://www.wto.org/english/tratop_e/sps_e/01_ms_camille_flechet.pdf

Yadavajitrikamji, A. (ed.) (2008) Vridhha Sushruta: Acharya Sushruta, Nagarjuna, Chandrat, *Susruta Saṃhita*, with Commentary Nibandhsangraha of Dalhanacharya, Reprint. Chaukhambha Surbharti Prakashan, Varanasi.

Zahedi, S.G., Qorbani, M., Nasli-Esfahani, E., Aghasi, M., Khoshamal, H. *et al.* (2021) The effects of *Elettaria cardamom* supplementation on inflammatory markers and vascular function in patients with type 2 diabetes mellitus: A mechanism-based randomized clinical trial. *Journal of Herbal Medicine* 25, 100403. DOI: 10.1016/j.hermed.2020.100403.

Printed and bound by CPI Group (UK) Ltd, Croydon, CR0 4YY

14/09/2026

14961845-0001